EIGHT YEARS TO
THE MOON

THE HISTORY OF THE APOLLO MISSIONS

NANCY ATKINSON

AUTHOR OF *INCREDIBLE STORIES FROM SPACE*

FOREWORD BY RUSSELL SCHWEICKART, APOLLO 9 ASTRONAUT

PAGE STREET
PUBLISHING CO.

PAGE STREET
PUBLISHING CO.

First published in 2019 by

Page Street Publishing Co.

27 Congress Street, Suite 105

Salem, MA 01970

www.pagestreetpublishing.com

Distributed by Macmillan, sales in Canada by The Canadian Manda Group.

23   22   21   20   19   1   2   3   4   5

ISBN-13: 978-1-62414-490-5

ISBN-10: 1-62414-490-X

Library of Congress Control Number: 2019936042

Cover and book design by Laura Gallant for Page Street Publishing Co.

Photography by NASA, Draper, Norman Chaffee, Phil Covington, Ken Goodwin, Gary Johnson, C. Mac Jones, Earle Kyle, Rich Manley, Harold Miller, David Moore, John Painter and Carl Shelley.

Cover image © National Geographic Image Collection / Alamy Stock Photo / O. Louis Mazzatenta

Printed and bound in the United States

TO RICK

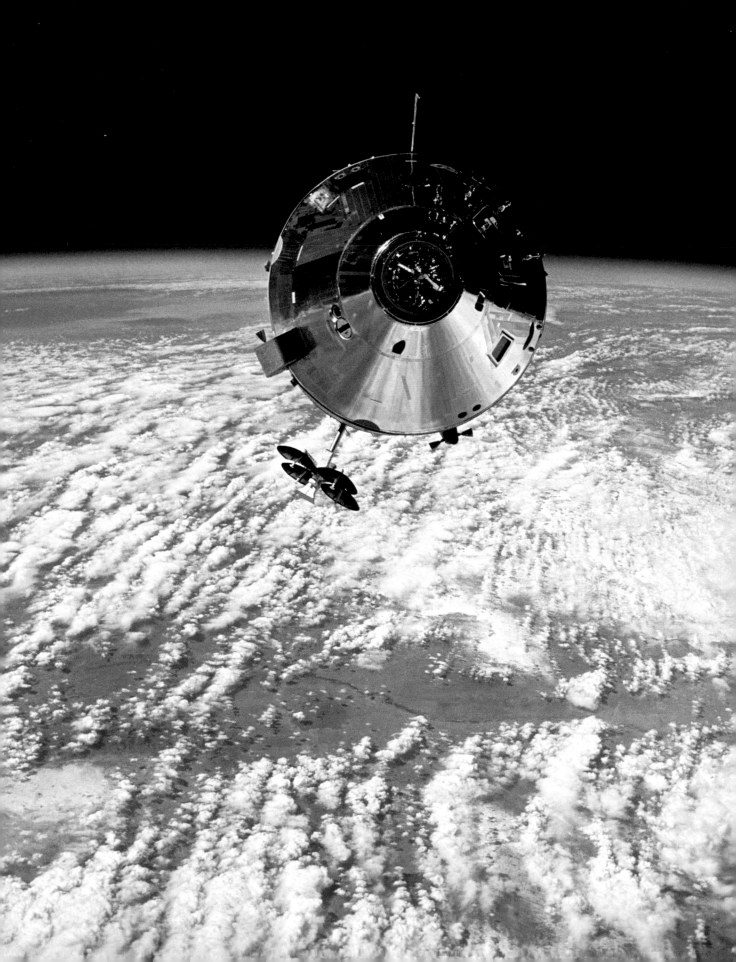

TABLE OF
CONTENTS

Left: The Apollo 9 Command Module in Earth orbit. Credit: NASA.

I'm a big-picture kind of guy. Since I've had many opportunities recently to reflect on the Apollo program—now that 50 years have gone by—there is one big-picture aspect of those days in the 1960s and early 70s that clearly stands out for me; to wit, the existential meaning of Apollo. What Apollo really gave us was a different perspective on the human condition.

The key moment for me was Apollo 8 in 1968. As the crew flew towards the Moon, they watched the Earth get smaller and smaller as it occasionally drifted through their windows. But their mission objective was to focus on the Moon. So when they went into lunar orbit, that's what they did. For their first three orbits they looked down at the Moon, studying the gray, cratered and lifeless lunar surface. As Bill Anders once said, "You see a thousand craters and it starts to get boring; one crater starts to look like another."

On their fourth time coming around the Moon, Commander Frank Borman rolled the spacecraft around the vertical so that the windows faced towards the forward horizon. Suddenly, this beautiful blue and white ball rose up over the colorless Moon. The shock of what they saw, the beauty and wonder of this stunning blue and white planet, contrasting so starkly with the devastated surface of the Moon and the ultimate blackness of space made them realize the significance of life and of home.

While this happened for the first time to three people named Borman, Anders and Lovell, in reality, it happened to all of us. For me, that is the significant thing about Apollo: For the first time, we had the ability to understand the preciousness and uniqueness of life; to realize that we are the only life in this little corner of the universe. And now, after 50 years, we've come to understand that we can be an extremely strong and powerful influence on where the evolution of human life goes from here. But with power comes responsibility.

Less than three months later, during my own Apollo 9 mission, I had, by happenstance, a five-minute break while floating outside our Lunar Module. I proactively opted to step out of my astronaut role and fully take in where I was and what I was seeing. Uninvited, a flood of questions popped into my mind; How did I get here? What's really going on here? Who am I? What does "I" mean? Am I "I" or am I "us"? These were not small questions and I wrestle with them still after 50 years.

So for me, here on its 50th anniversary, it is these personal experiences, gifted to a few lucky astronauts directly but to all of us in reality, that shape the historic significance of Apollo. That said, the other essential reality of Apollo is that it took over 400,000 people to make it possible to get to the Moon. For those of us who lived it day to day, we worked directly with perhaps a few handfuls of people, and we ended up meeting maybe a thousand or so others as we traveled across the country, visiting the sites where people were doing essential things for Apollo. But there were thousands of others who comprised this incredible tree of 400,000. Each and every person on that tree experienced a life-shaping, life-changing story of the dedication it took to make Apollo possible.

What Nancy Atkinson does with this book is pick up the behind-the-scenes details to bring home the fact that many real and ordinary people came together to do this amazing, extraordinary thing. Nancy shares their stories in a way that highlights how all these people were living and breathing the responsibility of contributing to humanity's dream to reach the stars. She makes this history more intimate, and shows us how everyone who worked on Apollo was an essential part of "one small step for man, one giant leap for mankind." It's a moment that will never be repeated and should never be forgotten.

I'm delighted Nancy decided to tell the Apollo story in this way. While the stories in *Eight Years to the Moon* are just a sampling of the 400,000 stories that are out there, this sampling comes at a deeper level that has not generally been heard, and provides an intuitive view of those who worked on the myriad bits and pieces of Apollo. Apollo was a tremendous collective experience of discovery and creativity and accomplishment. And that's a great story.

I am personally indebted and thankful to every one of those 400,000 people who each contributed, in their own way, not only to meeting JFK's goal "of landing a man on the Moon and returning him safely to the Earth," but to having played a personal role in realizing this historic shift in humanity's understanding of its place in space and time.

SLEEPING ON THE MOON

NEIL ARMSTRONG AND BUZZ ALDRIN WOKE
up on the Moon with two big problems.

Woke up, however, is perhaps a bit of an overstatement.

Neither astronaut was able to fall asleep after everything that had just happened. After all, it was a day of history-making and drama unlike any other: humankind's first landing on the Moon, our first foray on another world. And getting down to the lunar surface wasn't easy. The landing was fraught with communication problems, electrical glitches and navigation errors so great the Lunar Module (LM) overshot the intended landing site by more than four miles. Then Armstrong had to search for a new place to set down that wasn't filled with treacherous, LM-tipping boulders, manually flying the lander until it nearly ran out of fuel. All the while, jarring alarm klaxons sounded as the overloaded navigation computer struggled to process excessive data.

Finally, though, they landed—successfully. And all those in the Mission Control Operations Center (more commonly known simply as Mission Control) in Houston who had nearly turned blue from holding their breath started breathing again.

"Be advised there're lots of smiling faces in this room, and all over the world," Capcom and fellow astronaut Charlie Duke radioed up to the crew on the Moon.

"Well, there are two of them up here too," Armstrong replied.

"And don't forget one in the command module!" chimed in crewmate Mike Collins, orbiting above the Moon in the *Columbia* Command Module. Collins had also been holding his breath while listening in on the communications loop— and holding a hopeful vigil in the crew's ride back to Earth.

The Apollo 11 astronauts' view of Earthrise, from lunar orbit. Credit: NASA.

The original mission plan called for a sleep period for the astronauts after the landing and before going out and doing the historic moonwalk, or the extravehicular activity (EVA) in NASA parlance. But who could sleep after what they'd just been through? Armstrong and Aldrin petitioned Mission Control—and the EVA was moved up, right into prime time on a Sunday night back in the US (or, depending on whom you talk to, this was the secret plan all along).

Around the world, approximately six hundred million people, or one-fifth of humanity at the time, huddled around small television sets in homes and bars and on street corners, watching in wonder at the fuzzy black-and-white transmissions from the Moon. They saw the ghostly visages of two terrestrials bounding about on an alien world. For two hours and thirty-two minutes, the astronaut duo put their bootprints on the Moon, collected rocks, planted a flag, set up several science instruments and even talked to the president.

And *then* it was time to get some shut-eye. The two astronauts had been awake for more than twenty-two hours.

Neil Armstrong inside the Lunar Module Eagle *after the historic first moonwalk. Credit: NASA.*

While every system on the LM had been checked and rechecked, verified and signed off on, for some reason no one had spent much time considering where and how the first two humans on the Moon would sleep. Subsequent Apollo crews would have hammocks that could be strung up inside the cramped quarters, but Armstrong and Aldrin had to make do by trying to curl up in the limited flat areas. The floor space where the astronauts stood while landing their LM, the *Eagle*, was just 36 by 55 inches (1 by 1.4 m). The only other area suitable to use for sleeping was a ledge that covered the LM's ascent engine—the engine that would blast them off the Moon—but that space wasn't very wide or long. So, while Aldrin lay in a semi-fetal position on the floor, Armstrong tried to stretch out on the ascent engine cover, his legs in a makeshift sling fashioned from a tether to keep his feet level with his torso.

Outside, the Moon was silent, but the LM wasn't. The *Eagle* creaked from temperature changes on its metal exterior, with a nearly 500°F (260°C) difference between sun and shadow on the airless Moon. Inside, the tiny cabin was noisy with various systems—especially that all-important life-support system—running in the background.

"There was this pump," Armstrong would recall later. "My head was back to the rear of the cabin and there was a glycol pump or a water pump or something very close to where my head was." If Armstrong started to doze, the pump would kick in at an inopportune time and wake him.

In addition, between the bright panel display lights that couldn't be dimmed and the sunlight streaming into the windows, it was virtually broad daylight inside the *Eagle*. The astronauts had pulled down the window shades, but that didn't help. Armstrong described the shades as having the light-dimming qualities of photograph negatives.

"The light came through those window shades like crazy, and they were glowing, illuminated by intense sunshine outside," he said. "Then after I got into my sleep stage and all settled down, I realized there was something else."

The *Eagle* had an optical telescope sticking out like a periscope, which the crew used as a sextant in tracking the stars for navigation, just like the early ship navigators on the Earth's seas. "Earth was shining right through the telescope into my eye. It was like a light bulb," said Armstrong. He got up and hung an extra moonrock bag over the eyepiece.

Armstrong was congested, probably a reaction to the lunar dust. He and Aldrin had tracked plenty of sootlike lunar regolith into the spacecraft, and it seemed to magically stick to their suits, gloves and everything else like static cling.

Armstrong and Aldrin tried closing their helmets to shut out the noise and light. They also used their suits' ventilation system for better air quality. But that didn't work either. The suits' cooling systems—necessary for regulating body temperature on the scorching lunar surface—was working a little *too* well inside the cold cabin. It was only about 62°F (17°C) inside the *Eagle*, and both astronauts were cold. They finally turned off their cooling and ventilation systems, opened their helmets and made a mental note to suggest blankets for future Apollo crews.

The best Aldrin could manage was a "couple hours of mentally fitful drowsing." Armstrong just stayed awake.

Back in Mission Control in Houston, NASA flight surgeon Dr. Kenneth Beers was monitoring data from the astronauts' biomedical sensors. He watched as Armstrong's heart rate would start to go down into the sleep range only to go back up again, indicating he was stirring around considerably.

"This was a never-ending battle to obtain just a minimum level of sleeping conditions, and we never did," Armstrong said later during a mission debriefing. "Even if we would have, I'm not sure I would have gone to sleep."

Armstrong's statement was understandable: Besides the lack of creature comforts keeping him awake, there was the simple fact that he and Aldrin were on the Moon.

They were on the freakin' Moon.

Panoramic view of the Apollo 11 landing site. Credit: NASA.

Throughout all of human history—up to 1969, anyway—a trip to the Moon was considered far-fetched if not downright impossible. Even once it became theoretically possible, few believed that it could successfully be accomplished. But now Armstrong, Aldrin and Collins had done it. They did it not for themselves, but for the hundreds of thousands of people who had worked their collective asses off for the past eight years, enduring a lot of personal sacrifice, to make it all possible. Eight years of working late nights and weekends, missing their toddler's first steps, missing their kids' ball games and first dances and maybe even growing apart from their spouse. "A lot of us who were married during that time ended up not married later," said one Apollo engineer.

But they did it. They did it amid the tumultuous times of the 1960s: civil rights, the Cold War, the Vietnam War, the free-speech movement, the women's movement, hippies and the assassinations of Jack and Bobby Kennedy and Martin Luther King Jr. And now, if only for a few days, everyone could put their differences aside and celebrate the success of this historic moment.

And they did it in a way that made the event belong to all of humanity—as the plaque left on the Moon by the Apollo 11 astronauts reminds us, "We came in peace for all mankind." Surely, if we can send humans to the Moon, it would seem we might be capable of just about anything.

And so, as they tried to sleep on the Moon, all Armstrong and Aldrin could do was lie there, probably thinking about the magnitude of what they had just accomplished. But in all likelihood, what really kept them awake was what loomed ahead of them. It was the other side of the equation that President Kennedy had challenged NASA to do eight years earlier in 1961: to not only send men to the Moon but also to return them safely to Earth.

And therein lay their two problems.

A replica of the plaque the Apollo 11 astronauts left behind on the Moon in commemoration of the historic mission. Credit: NASA.

The first issue was that no one—neither on Earth or on the Moon—knew where the *Eagle* had landed. Yes, they were on the Moon, but precisely *where* they were had ramifications for the upcoming lift-off and rendezvous with *Columbia*.

The navigation computer's errors and Armstrong's fast-on-his-feet flying to find a safe landing spot meant that the *Eagle* wasn't where everyone expected it to be. Issues with the radar and communications during landing made it difficult to get an exact fix on their location.

Flight dynamics officer Dave Reed in Mission Control was monitoring the situation. He took his headset off so as not to be overheard, walked back to flight director Gene Kranz and said, "We have a problem: We do not know where the hell they are."

The view out the window of the Lunar Module of the Sea of Tranquility on the Moon. Credit: NASA.

In spaceflight, timing is everything. Knowing the *Eagle*'s exact location would establish the timing of lift-off from the Moon in order to meet with *Columbia* in lunar orbit. For the Apollo program, the time at which the LM took off from the surface of the Moon would set up a sequence of events—from rendezvous and docking with *Columbia*, to the time of leaving lunar orbit, all the way to coordinating where and when in the Pacific Ocean the aircraft carrier USS *Hornet* would have to be to pluck the crew from the Command Module after splashdown. Change the lift-off time, and the *Hornet* would have to steam across the ocean to a different spot.

Armstrong and Aldrin tried to give landmarks that might provide clues, but there wasn't a lot to go on. Out the window, the astronauts could see a relatively level plain, with a large number of craters and ridges and a hill that might be a half mile away. Or was it farther? Judging distances without any frame of reference was extremely difficult. Plus, those features would describe just about any area of this lunar plain called the Sea of Tranquility. Armstrong mentioned the football field–size crater strewn with boulders they had flown past.

Collins, up above in *Columbia*, was enlisted to help. As he circled the Moon every hour and fifty-eight minutes, he scanned a grid of possible coordinates radioed up from Mission Control, his eye to the Command Module's twenty-eight-power sextant. But whizzing along at 3,700 miles per hour (6,000 km/h) and just 69 miles (111 km) up, the landscape below zipped by and Collins had—at most—about two minutes to frantically search the grid. Sometimes he intently scanned the landscape with his eyes alone, trying to catch at least a glint off the LM or some other clue to visually pinpoint the *Eagle*'s landing site.

Every time Collins came around from the far side of the Moon, Mission Control would have a new set of coordinates for him to scan.

"I need a very precise position, because I can only do a decent job of scanning maybe one of those grid squares at a time," he radioed back to Earth, a sense of frustration in his voice. "The area that we've been sweeping covers tens and twenties and thirties of them."

Before Collins could bed down for the night, *Columbia* made one more pass over the area the flight controllers thought might be *Eagle*'s landing site. Again, he squinted, scanning the horizon below.

"No joy," Collins finally said.

While the crew tried to sleep, the flight controllers back on Earth would have to come up with another plan to try to pinpoint *Eagle*'s position.

The second problem was of more significant concern.

As Aldrin got down on the floor to sleep, he saw something on the floor. He reached over, grabbed it and found it was a broken-off switch of a circuit breaker. He looked up to the panel of knobs and switches that ran his flying machine and saw that it wasn't just any old circuit breaker. It was the engine-arm circuit breaker, the switch that activated the ascent engine that would lift them off the Moon to rendezvous with Collins.

Aldrin realized that since the panel was on his side of the spacecraft, he must have knocked it off while moving around in his spacesuit, bumping it with the huge backpack-like Portable Life Support System.

He radioed to Mission Control.

"Houston, Tranquility. Do you have a way of showing the configuration of the engine-arm circuit breaker? Over." He paused. "The reason I'm asking is because the end of it appears to be broken off. I think we can push it back in again. I'm not sure we could pull it out if we pushed it in, though. Over."

The engine-arm circuit breaker would have to be switched at just the right moment to send electricity to turn on the ascent engine.

Capcom Bruce McCandless radioed back: "Roger. We copy. Stand by, please." After the flight controllers checked their readouts, McCandless came back on.

Astronaut Buzz Aldrin inside the LM during the Apollo 11 mission. Credit: NASA.

"Tranquility Base, this is Houston. Our telemetry shows the engine-arm circuit breaker in the open position at the present time. We want you to leave it open until it is nominally scheduled to be pushed in, which is later on. Over."

So while the astronauts slept—or tried to sleep—engineers back on Earth had to figure out a new plan. If the crew couldn't get that breaker to work, they'd have to figure something else out or there'd be no ascent. Aldrin and Armstrong would be trapped on the Moon.

WHILE THERE WERE THREE PEOPLE ON or orbiting above the Moon, it took approximately four hundred thousand people from across the US and around the world to make this first Moon landing possible. From the dreamers who thought it could be done; to the engineers who worked out all the details; to the workers who torqued the wrenches on the Saturn V rocket, *Columbia* and *Eagle*; to the seamstresses who sewed the spacesuits; to the computer programmers who punched the code; to the engineers who designed and built all the systems; to the scientists, the trainers, the navigators, the technicians, the managers, the secretaries, the supervisors, the flight controllers and flight directors.

Armstrong never missed an opportunity to praise the hundreds of thousands of people who made Apollo possible. He once said, "Every person, man or woman said, 'If anything goes wrong here, it's not going to be my fault.'"

And so, as Armstrong and Aldrin lay sleepless on the Moon, back on Earth there were four hundred thousand others who weren't sleeping either. They were either working feverishly in Mission Control, or in one of the backroom support teams in Houston to ensure the crew could leave the Moon and return home, or at one of the thousands of contractor company sites across the country, crunching the numbers to make sure all the hardware and software was going to work as advertised. Or they were in their own beds, tossing and turning, wondering if they had risen to the challenge and had done everything they could to ensure safety and success for the crew of Apollo 11 in this most hazardous and dangerous and greatest adventure on which humankind had ever embarked.

CHAPTER 1

1962

You learned to keep your pencil sharp.

—CATHERINE OSGOOD, engineer, Rendezvous Analysis Branch

WHEN KEN YOUNG ARRIVED IN HOUSTON

in June 1962, the first thing he did was drive about 25 miles farther southeast to the site where the new Manned Spacecraft Center (MSC) was going to be built. And all he found was cows.

What would eventually become a six-lane highway called NASA Parkway was at that time just a narrow oyster-shell road that stretched from the Webster railroad tracks to Seabrook, following the curve around Clear Lake. All along was open coastal prairie pastureland with grazing Herefords, longhorns and shorthorn Durhams. Young noticed a livestock water tank with a windmill whirring nearby.

The land was part of the 20,000-acre West Ranch, owned by the heirs of the Humble Oil and Refining Company, later known as the Exxon Corporation. The Wests had donated a 1,000-acre portion of their ranch to nearby Rice University, who in turn had offered the property to the NASA Space Task Group. This was a group of engineers that managed America's human spaceflight programs and was charged with finding a suitable location for a new complex of research labs, office buildings and test and control facilities so that NASA could send humans to the Moon with Project Apollo.

Left: Failed launch of the Mercury-Redstone 1 on Nov. 21, 1960. An electrical fault caused the rocket engine to shut down after the rocket rose just 4 inches (10 cm), triggering the Mercury capsule's escape rocket to jettison. Credit: NASA.

MSC SITE JANUARY 1962

A view of the Manned Spacecraft Center site in January 1962, prior to ground breaking and the beginning of construction. Credit: NASA.

This rural property in Harris County, Texas, was appealing to the Space Task Group since it met several of the site requirements, including access to barge traffic through Clear Lake—just on the south side of the road—which in turn provided access to Galveston Bay and the Gulf. The land was close to Ellington Air Force Base, providing easy air access; it was near institutions of higher education (Rice University and the University of Houston), and the region had a moderate climate "permitting out-of-door work for most of the year," according to the wishes of the Space Task Group's site selection memorandum. Plus, it didn't hurt that Texas was home to several influential US congressmen, such as Speaker of the House Sam Rayburn and Albert Thomas—the man who had the power over the country's purse as chairman of the House Appropriations Committee—as well as Vice President Lyndon Johnson. They were all big supporters of the space program, especially with the economic benefits that a big, new and prestigious facility in their state would bring. Congress had just passed a $1.7 billion NASA appropriations bill that included $60 million for the new facility in Houston.

And so, in September 1961, NASA announced the Space Task Group's decision to build the new facility on this plot of land near Houston. From its inception, it was to be the lead center for all US space missions involving astronauts. The cows would have to go (although years later, they would return to graze once again in a special pasture set aside to educate tourists about the history of this place).

But in 1962, the MSC would soon be the place where people would design, develop, evaluate and test the spacecraft for Project Apollo (as well as all of its subsystems) and train the crews that would fly these missions. The ideas were there, the dreams were there—but just how to implement all these monumental tasks was mostly unknown. The primary need was people, and in particular, brain power. NASA would need to transform from a small research organization into a large federal agency teeming with scientists, engineers and managers all to figure out how to do things that had never been done before.

In that year alone, more than two thousand new hires came pouring into Houston. The incoming recruits had one thing in common: they were young, either fresh from college or the military or plucked from the oil, aircraft or electronics industries. Some were single, crew-cutted and wide-eyed—and when they weren't working, they were on the lookout for fun and adventure. Many were already married with young families, families that formed the basis of the close-knit communities that soon sprang up.

Ken Young was among some of the first new hires to show up, one of the near-originals—his number at the new credit union for NASA employees was 173. But the real originals were the one hundred or so folks who were part of the Space Task Group that had transferred in during the winter and spring of 1962 from the Langley Research Center in Virginia and the Lewis Research Center in Cleveland, Ohio. The group included thirty-seven engineers, eight secretaries and math aides (the women who did all the mathematical calculations and prepared the graphics) plus an additional thirty-two engineers from Canada who moved south after the Avro Arrow project—a specialized interceptor aircraft that was going to be constructed in cooperation with the US—was canceled.

NASA had given Young a pretty decent job offer, he thought. It wasn't the best he got, but he knew he could work at the new center in Houston and, being from Austin, he didn't want to leave Texas. He took it.

Most importantly, he was going to be working on something related to space—he knew that much—but as far as a particular task or job, he didn't have a clue. He checked in at the NASA personnel headquarters in a small upstairs office in the East End State Bank Building on Telephone Road in southeast Houston. After describing his interests and education with a personnel manager named Leslie Sullivan, Young was placed with the Mission Planning and Analysis Division. He was going to be working on figuring out trajectories for launch, orbit and reentry. One other aspect intrigued him: the rendezvous of two spacecraft, which was one of those things that had never been done before.

"All I knew was that I wanted to work on trajectories and orbits and stuff, but I went in there with no real idea," Young said. "There weren't any textbooks on the subject yet, but my new manager, Bill Tindall, had compiled a manual that was called the Space Notes, and as new hires—they just hired a bunch of us—we had to sit there and memorize things from this 3-inch-thick, stapled-together handbook and solve equations and work out problems with our slide rules, just to learn the basics of orbital mechanics. Hardly anybody knew how to do anything."

Since construction of the MSC was just getting under way, everyone who came to Houston for NASA was put to work in an assortment of about fifteen different buildings on the southeast side of Houston, now the property of the US government, either through leases, purchases or appropriation because of back taxes. Young went out with his group to the old Houston Petroleum Building, which had the distinguishing feature of a rusty oil derrick out front.

But Young settled in, soaked up new information like a sponge and found a place to live. At the end of his first week, he went to the credit union to borrow $200 so he could buy a black-and-white television set for his apartment. A guy with a new job had to have at least one small luxury.

Young had just graduated from the University of Texas in Austin. When he enrolled there, his original plan was to be an engineer—whatever the hell that was, it sure sounded good. But he was always analyzing things, interested in how different contraptions worked, and excelled in math. When he found out that civil engineers built bridges and such, he was intrigued.

But just two weeks after classes started his freshman year, all that changed. It was a Friday—October 4, 1957—when the Soviet Union launched *Sputnik 1*, the world's first artificial satellite. The small spacecraft, 23 inches (58 cm) in diameter, and its booster stage could be seen from Earth at dusk, orbiting the planet once every ninety-six minutes. Its eerily repeating *beep-BEEP-beep-beep*, *beep-BEEP-beep-beep* could be heard by amateur ham radios and was broadcast by radio stations as it continually transmitted signals back to Earth.

That launch a half a world away changed the direction of Americans' lives. First thing Monday morning following *Sputnik 1*'s launch, Young marched over to the university's registration office and changed his major to aeronautical engineering. There wasn't anything called astronautical or aerospace engineering yet, but Young knew he wanted to be involved in this business of going to space. And he wasn't the only one.

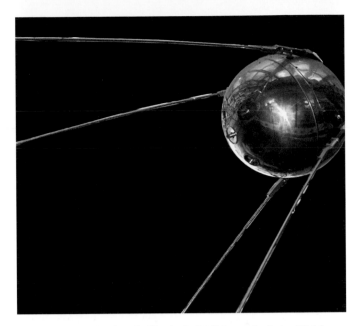

A replica of *Sputnik 1*, launched by the Soviet Union as the first artificial satellite to be launched into space. This replica is stored in the National Air and Space Museum. Credit: NASA Space Science Data Coordinated Archive.

Sputnik 1 was launched during the International Geophysical Year, a year dedicated to worldwide research on satellites and Earth's atmosphere. The international scientific community had a goal of launching a satellite to orbit Earth, and there was a race to see who could do it first. Russia's surprise launch of *Sputnik 1* fueled the space race between the Soviet Union and the United States, intensifying the tensions between the two countries locked in the Cold War.

In the US, some people were in awe that humans could actually launch an object beyond the bounds of Earth and watched in wonder as the 184-pound (83-kg) sphere passed overhead. But mostly, a profound shudder went through the populace. Many were worried about falling behind "the Communists," that our education system wasn't good enough, that we had failed to encourage research and technology development, that our government was made up of a bunch of slackers. Others were just plain frightened. If the Soviets had rockets capable of sending objects to space, could they launch nuclear weapons to North America? Everything about *Sputnik* seemed threatening.

It didn't help that a month later the USSR launched *Sputnik 2*, with a dog named Laika inside a 1,100-pound (500-kg) payload. Putting a critter into space likely meant the USSR's goal was to soon launch humans as well, and the US was nowhere near that possibility.

"When I saw the dog go up, I said, 'My God, we better get going because it's going to be a legitimate program to put man in space,'" said Robert Gilruth, who at that time led teams of engineers at NASA's originating institution, the National Advisory Committee for Aeronautics (NACA), and later became the leader of the Space Task Group. "I didn't need somebody to hit me on the head and tell me that."

Engineers at NACA had already been testing human capabilities and thresholds regarding spaceflight—in particular, g-forces—and they felt they understood some of the physiological limits well enough to start designing a human spacecraft. Then, through ballistic missile tests at a launch facility on Wallops Island, Virginia, they had data on how much heat was generated by high-velocity reentry through Earth's atmosphere. Their findings showed that a relatively new concept of using a preceding shock wave in front of a spacecraft was the best way to reenter a human-size vehicle through the atmosphere. That meant using a cone-shaped, blunt-ended spacecraft. It was a complete change in architecture to go from flying a pilot in a winged vehicle—which NACA had been advancing since 1915—to putting an astronaut lying on his back in a space *capsule*.

The capsule concept was an extremely hard sell for engineers who had worked on airplanes most of their careers. Plus, many space-enthused engineers and designers had been inspired by science fiction (with Buck Rogers's winged ships depicted in comics and pulp magazines) or the series of 1950s articles by rocket engineer Wernher von Braun in *Collier's* magazine with fantastic winged spaceships portrayed by artist Chesley Bonestell. Hugh Dryden, the head of Langley, equated the space capsule concept to "shooting a lady out of a cannon." Others at Langley weren't convinced this infatuation with space travel might not be much more than a circus stunt or a passing fancy.

Gilruth began working with leaders in Washington, DC, to determine how the United States could get a human to space. But while the US had been launching ballistic missiles and other rockets for years, actually getting a payload into Earth orbit with multistage launch vehicles proved much more difficult, even though different branches of the military were working on it—and each in their own way. In a fierce competition, the Army was trying to get their Jupiter and Juno rockets flying before the Air Force could get their Thor or Atlas launchers off the launchpad.

In December 1957, the US Naval Research Lab project attempted to launch its Vanguard rocket with a satellite that was about the size of a grapefruit and weighed 3 pounds (1.3 kg). It rose a few feet into the air before blowing up in a massive, spectacular fireball. The event was broadcast live on national television, and the US press dubbed it "Flopnik" and "Stayputnik." More remarkable launch failures followed while the Soviets continued with a series of successful missions. Americans were embarrassed.

Suddenly the ability to launch rockets became a national priority. The Army Ballistic Missile Agency—directed by the rescued World War II team of German rocket engineers led by von Braun—quickly pooled resources with the Jet Propulsion Laboratory (JPL) in California, putting together a four-stage Juno rocket with JPL's canister-shaped, 30-pound (14 kg) satellite called *Explorer 1*. It launched successfully on January 31, 1958, the first US satellite. Its single instrument sent back data about the radiation environment high above Earth's surface.

With the advent of television and mass communications, the world now watched as the space race between the US and the Soviet Union unfolded.

But the United States needed to formally organize its space program. The Eisenhower Administration created NASA in October 1958. The new organization combined a small nucleus of engineers inherited from NACA (as well as personnel transferred from the Vanguard and Juno programs) and added the teams of the Army's German rocket men, now at the newly formed Marshall Space Flight Center (MSFC) in Huntsville, Alabama.

Left: Crews prepare a Little Joe rocket for launch from Wallops Island in the early 1960s. Credit: NASA.

Alan Shepard launches on the United States' first human spaceflight, the Mercury-Redstone 3, on May 5, 1961. The suborbital mission attained a maximum speed of 5,180 miles per hour, reached an altitude of 116½ statute miles and landed 302 statute miles downrange from the launch site, Cape Canaveral, Florida. Credit: NASA.

A few launch successes came amid numerous failures while engineers tried to figure out how they could possibly put a human into space with the new Redstone rocket and a human-size, blunt-end capsule called Mercury, designed by the Space Task Group engineers. Even with successful launchings of monkeys on suborbital flights, more rocket failures ensued, and NASA wanted to be cautious before actually putting a person inside one of these unpredictable, catapulting bombs. But a group of seven astronauts had now been selected with much public fanfare, and they were undergoing training while the Mercury capsule was undergoing tests. And rockets kept failing or blowing up with an alarming frequency.

Members of NASA's Space Task Group in 1958. Credit: NASA.

Further embarrassment came during the first launch attempt for the Mercury-Redstone in November 1960. The rocket rose just 4 inches (10 cm) and then its engines shut down, slamming it back on the pad. The rocket didn't explode, but in an almost comical, confused afterthought, the escape tower shot up into the air, and the capsule's drogue parachute popped up like a cork and ended up draped innocuously along the side of the rocket. That may have been the absolute lowest point of morale among those who worked on Project Mercury.

Meanwhile, some members of the Space Task Group—always the innovators with a look toward the future—formed a Research Steering Committee on Manned Space Flight. Headed by Harry Goett, the committee started mapping out an assortment of long-range strategies for NASA. Some thought von Braun and his team had the perfect idea of building gigantic rockets to explore the solar system. Some scientists thought Earth satellites to monitor the environment were important. Several engineers looked into what it would take to go to the Moon, and in fact, the Goett committee recommended either a circumlunar trip or a manned lunar landing as appropriate long-term goals of NASA's space program. But first, they suggested a major interim program to develop advanced orbital capabilities, such as the construction of a space station. But this was all far into the future, with no meaningful (i.e., financial) support for any immediate effort for such missions. It was hard enough to launch small rockets with any sustained success, and new, young President John F. Kennedy had not given the issue of space much thought.

But April 12, 1961, transformed all that. The USSR launched cosmonaut Yuri Gagarin into a single orbit around the Earth. It was audacious and stunning, and Americans were shocked that the Soviets had won this race. The public, lawmakers and the media demanded quick action to counter the obvious fact that the United States was in second place on the space frontier.

Counteraction came just weeks later, when, on May 5, 1961, Alan Shepard became the first American to reach space, launching on a fifteen-minute suborbital Mercury flight. The success boosted US morale and strengthened hopes for the future of space travel. Shepard's admonition to the pad crew to "solve your little problem and light this candle" was indicative of the drive many Americans felt to get ahead of—or at least stay in step with—the Russians. Then suddenly, on May 25, 1961, just shy of three years after President Eisenhower signed an act creating NASA and just three weeks after Shepard's flight—meaning the US had only fifteen minutes of spaceflight experience—Kennedy set a goal for the United States that would surpass any previous engineering and scientific feat.

"I believe that this nation should commit itself to achieving the goal, before this decade is out, of landing a man on the Moon and returning him safely to the Earth," President John F. Kennedy told the American people in a speech addressing "urgent national needs" before a joint session of Congress. "No single space project in this period will be more impressive to mankind, or more important for the long-range exploration of space; and none will be so difficult or expensive to accomplish."

While most of the US public cheered this daring plan, there were NASA engineers who looked at each other and said, "What?"

NORMAN CHAFFEE WAS AT HIS HOUSE IN

Tulsa, Oklahoma, on the phone with a woman in the NASA office in the East End State Bank Building. She was looking at his résumé, and Chaffee was figuring out how he could convince her he would be a good fit for the new MSC.

Chaffee explained he was a chemical engineer about to get his master's degree and had some experience in the petrochemical industry but wanted to get into the space business. He'd always been a science fiction nut, spending any spare money he made from odd jobs in high school on books by Poul Anderson and Robert Heinlein, and he was enchanted with the idea of humans traveling to distant worlds in space. In the back of his mind, he always assumed those crazy ideas would remain permanently in the realm of science fiction. But suddenly the Russians accomplished Sputnik and the US was trying to keep up, and then President Kennedy announced the space program's goal to send humans to the Moon. That was all Chaffee needed to look into a career change.

Another enticement for Chaffee was that he could come back to Houston. He had spent three years at Rice during his academic days. He fondly recalled going down to the docks at Kemah with his college friends to pick up fresh fish. He loved pulling pranks on his dorm mates, and he even considered the hazing he received as a freshman as great fun. To him, the Houston area had a confluence of wonderful characteristics: It was warm, not far from the beach and fresh shrimp were easy to find. Because he was so excited about the whole concept, Chaffee wondered if the NASA organization was going to need chemical engineers.

"Well, I don't know, let me look at this staffing book that I've got," said the woman on the other end of the line, pausing. "Oh, yes, we apparently are looking for engineers, but we're mostly looking for mechanical engineers, electrical engineers and aeronautical engineers. But here is an area called Energy Systems."

"Do you know what that is?" Chaffee asked her.

"No, I really don't. It's a very generic description."

Chaffee thought for a moment. He'd been at the university level for seven years, taken courses in advanced thermodynamics, heat transfer and heater design and had dabbled in electrical engineering. Whatever Energy Systems was, he must at least have an inkling of it.

"Well, I just happen to be an expert in energy systems," Chaffee told her.

"Oh, well, my goodness, let me pass along your file and we'll let you know."

That conversation occurred in February 1962. A few weeks later, Chaffee got an offer from NASA by mail and a call from a fellow named Dick Ferguson, a branch chief from the Space Task Group who had moved to Houston. Ferguson offered Chaffee roughly the same amount of money that the oil refinery was paying him, but there were all the intangibles of doing something big and important for the country. It would be a dream job, Chaffee thought. He conferred with his wife, Olga, who had just given birth to a baby girl. "You just decide where you want to go, and we're going with you," she said.

Chaffee finished up his master's thesis, packed the car and the family and arrived for his first day of work on May 13, 1962.

Norman Chaffee in the mid-1960s. Photo courtesy of Norman Chaffee.

It turned out the Energy Systems Branch had responsibility for things like the pyrotechnics that powered the small steering propulsion systems on the spacecraft and power systems such as fuel cells and batteries—but not the big rockets that would blast astronauts into space, as Chaffee first assumed. Instead, he would be working on figuring out how to delicately and precisely steer a spacecraft from Earth to the Moon and back again. Chaffee was placed with the Auxiliary Propulsion Group of the Energy Systems Branch, led by Henry Pohl.

As a young man, Pohl had been in the army and worked with von Braun and his team in the 1950s building and testing rockets in Huntsville, Alabama. On Pohl's first day there, they told him to put on a hard hat and sent him out to a blockhouse to take part in the test launch of a Redstone rocket.

"When that thing lit off, I had never seen such power in my life," Pohl recalled. "That little blockhouse just shook. I decided right then and there that's what I wanted to be part of. I would have gladly given them all my $75-a-month pay grade to work on that thing."

Pohl helped invent new rocket ignition systems, designed the roll control systems used on the Jupiter launchers and helped solve some of the problems of the early rockets. Pohl recalled that the criterion for the launch to be successful back then was the rocket getting out of sight before blowing up. And a lot of them didn't make it that far.

Rockets became Pohl's passion: All he thought about was building rockets; all he dreamed about was building rockets. Soon he was helping design and test even more launchers: the Polaris, the Atlas, and then the first big rocket, the Saturn 1. They'd use wind tunnels, vacuum chambers, test stands—whatever facilities they had access to.

"You did everything," Pohl said. "I remember sketching out on a piece of notebook paper changes for an injector, taking it down to the shop on the way home at night, stopping by the shop the next morning, picking up that part, bringing it out to the test stand and having the technicians test that change that day. Look at the data that evening, make more changes, take it back to the shop that night, and have the changes made and modifications made, and bring it back and test it the next day."

Everything was moving fast and there were always problems piling up with never enough time to figure them all out.

"We didn't have a lot of supervision, we didn't have a lot of people telling us we couldn't do this, we couldn't do that, or you've got to do it this way," Pohl recalled, "and I think that's why we made a lot of progress. I was just always proud when the dadgummed things worked."

Every twist and turn, every failure and every success was a chance to learn, and in Pohl's mind, it was all fun. He always liked working hands-on to solve engineering problems and never wanted to be a supervisor. But all of a sudden, he was asked to move to a management role to help oversee all the working parts and people for the Redstone rocket for Project Mercury.

Even with Shepard's successful flight, Pohl knew intimately the variety of problems the US space program faced. And so, when he heard Kennedy's announcement of going to the Moon, he was dumbfounded.

"That was the stupidest thing I'd ever heard in my life," Pohl said in his Texas drawl. "I mean, you have to appreciate what we were working with in that day and time. We still had the vacuum-tube technology. Transistors were just coming into being. The Atlas was the biggest operating rocket that we had, and we were still having failures in seven out of ten flights with it."

But if the Moon was where the US space program was going, Pohl knew he wanted to help get it there. Now, in early 1962, he found himself landing in Houston—not far from where he was born and raised in south Texas—tasked with helping manage an even bigger project.

As passionate as he was about rockets and sending humans to space, he almost turned around and went back to Huntsville after the first week in Houston.

"We had these staff meetings that went interminably long into the evening," Pohl fumed, "and all they talked about was editing reports, who was late, who had parked in someone else's spot. Here we were trying to get to the Moon in eight years and all they did was talk about all this crap."

Relief came as the new, young and enthusiastic engineers began showing up in Houston. Pohl immediately liked Chaffee.

"Norm, we are going to put you to work building rocket engines," he explained to the new guy.

Now, with his move to Houston, Pohl was asked to take everything he had learned, invented, tested and built on the big rocket engines and miniaturize it. NASA needed him to apply his knowledge and expertise to the little steering rockets for the Gemini and Apollo spacecraft so they could maneuver in space.

"And Mr. Chaffee, what we need you to figure out is going to be a completely different device than what we've built before," Pohl started out. "And no, we are not working on the big ones that launch you into orbit or take you to the moon. No, no. Those engines come on once and burn for eight minutes or so, and then they are done. What you and I are working on are small, low-thrust kinds of things that can fire in short, hot pulses, maybe thousands of times during a mission and get us to move about wherever we want in space. And that's what we want you to work on."

"Well, okay. That sounds like it should be pretty interesting," Chaffee replied slowly, trying to recall the advanced physical chemistry class where he had worked on exactly one homework problem that dealt with the thermodynamics of a rocket engine. Up to this point, that was his only experience. But he was about to become an expert among just a handful of people in the entire world who knew anything about these little rockets called the Reaction Control System (RCS). Chaffee was a tabula rasa who flourished under the tutelage of Pohl and others around him.

He got stacks of books and manuals to read, studied schematics and drawings, looked at reports and technical papers and found his old physical chemistry textbook that had been packed away in a moving box. He immersed himself in what he considered a wonderful and fortuitous assignment that—for a chemical engineer—was incredibly interesting and unbelievably exciting.

Then there was actual flight data to study from the Mercury flights: Gus Grissom had followed Shepard's *Freedom 7* mission with another ballistic flight on a Redstone rocket in July 1961 (*Liberty Bell 7*), and then John Glenn's historic, nail-biting three-orbit *Friendship 7* flight using the Atlas rocket came in February 1962. Scott Carpenter's Mercury-Atlas *Aurora 7* flight was in May 1962, right after Chaffee started work at NASA. By looking at the postflight data, he was coming to understand what these tiny propulsion systems entailed.

But, as Pohl had explained, the Gemini and Apollo RCS was going to be fundamentally different from what Mercury was using. Chaffee quickly came to understand that all spacecraft needed a system of small, low-thrust rockets that fire in extremely short pulses—just milliseconds worth of little pops—to steer, point or hold steady the attitude of the spacecraft, both during flight and reentry through Earth's atmosphere.

The new thrusters Chaffee was working on had to be larger and use fuels that were hypergolic, meaning that when they were combined, the fuel and an oxidizer would ignite and burn on contact and didn't need an ignition source like a spark or flame. That concept simplified the design of the RCS engines considerably. But a challenge arose in designing the part of the engines called the injector, which Chaffee saw as akin to the showerhead in his bathroom. It squirted out tiny amounts of the fuel and oxidizer in little streams into the part of the rocket engine called the combustion chamber so the two materials could ignite. The hot gas—about 5,400°F (3,000°C)—created by the ignition shot through an area called a throat, which squeezed down the gas and forced it to accelerate out through the nozzle, creating the thrust of the little rocket engines.

APOLLO COMMAND AND SERVICE MODULES ENGINE LOCATIONS

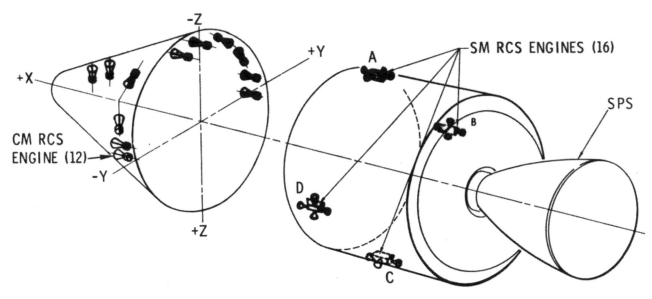

Technical drawing of the Apollo Command and Service Modules thruster and engine locations. Credit: NASA.

Astronauts review a map of the orbital track for a Gemini mission in 1965. Shown (left to right) are astronaut John Young, Donald Slayton, assistant director for Flight Crew Operations, astronaut Virgil (Gus) Grissom and Ken Nagler, US Weather Bureau. Credit: NASA/Johnson Space Center (JSC).

This whole series of violent events for a single ignition needed to happen in about twenty milliseconds, meaning you could fire the engine twenty times a second if needed.

Chaffee learned that if he could design it just right, the injector would distribute the fuel and oxidizer in just such a way as to create the most efficient combustion and not ignite in a way that would damage the interior of the combustion chamber. His chemical engineering background was beneficial, but he learned a tremendous amount in a short period of time through hands-on work.

After a couple of months, Pohl wanted to broaden Chaffee's experiences and have him sit in on a meeting with people from across the space agency in which issues on various systems for the Gemini spacecraft were being discussed.

"Norm, there's a meeting at the Gemini program office," Pohl said, explaining that Chaffee would have to drive over to the Veterans Administration building in downtown Houston since the various offices were still spread out all over town. "Item number eight on the agenda is something about the RCS engine, and we want you to go down and just take notes. You don't know enough yet to really contribute, but go down and see what the issue is and bring back the report."

Chaffee arrived at the meeting, took an inconspicuous seat and listened intently to try to learn as much as he could. He noticed astronaut John Young was in attendance. Young was among the second group of astronauts—the "New Nine"—selected by NASA to augment the original Mercury 7 for the upcoming Gemini and Apollo missions, and he had just been slated for the first Gemini flight, along with Gus Grissom.

After sitting through a few discussions he didn't entirely understand, Chaffee looked down at the agenda sheet he'd been handed and saw the next item was "Reliability of Zippers in the Gemini Spacesuit." A gentleman from the contractor constructing the suits, the David Clark Company in Worcester, Massachusetts, explained that at this time they couldn't guarantee the reliability of the zippers. The issue was that several of the specialized pressure zippers had been added to the suit, and there was a precise calculation for the reliability against leakage based on each linear inch of zipper. The required level of reliability was no longer being met because there were just too many linear inches of zippers. He recommended the 12-inch (30-cm) crotch zipper be eliminated, and that would solve the problem.

Since the Gemini missions were going to be several days long—perhaps up to fourteen days—the subject of the astronauts having to defecate during the mission was a known issue that was being considered by engineers and scientists. While urine collection bags were based on the time-honored "motorman's friend" with a hose and bag, the poop problem hadn't yet been completely solved. NASA dieticians were working on low-residue diets designed to minimize bowel movements, and other people were working on developing a fecal containment system—basically a plastic bag to be inserted through the crotch zipper, with adhesive around the opening to stick to the proper body area. The baggie would be accompanied by chemically enhanced wipes to kill bacteria and neutralize odors. But all of this was complicated by the issue of zero gravity—body wastes tended to either stick to the astronaut's behind or float around. And since the Gemini spacecraft was about the size of the front seat of a small car, this zipper issue was going to be an important decision because the other option was having the astronauts wear some sort of diaper that had a self-contained area inside the suit for stowing fecal matter.

John Young jumped up and went berserk. He was furious. He made an impassioned speech about being able to make better zippers and doing better for the program and the country. His bottom line was, "You've got to do better, and if you can't do better, then we're still just going to have a crotch zipper because I'd rather suffocate than crap in my pants."

At the end of the day, the decision was made to find better zippers.

When Chaffee returned to the office, Pohl greeted him and asked how the meeting went and if he had learned anything.

"Yes," Chaffee said. "I learned that John Young is not going to crap his pants."

NOT EVERYONE WAS COMPLETELY TAKEN

off guard by Kennedy's announcement that NASA would reach for the Moon. Some who worked in the trenches of the space program knew what might be possible. Chester Vaughan had been with NACA at Langley since 1955, and he saw the progression of technology and the paradigm shifts in the efforts toward spaceflight.

"All the work that was done earlier allowed us to be in the position so when Kennedy announced the eight years of allowable time to get to the Moon, we already had a healthy head start," Vaughan said. "Not that everything was defined directly, but we had a good feel for what had to be done."

Vaughan started working for NACA through a co-op work training program at Virginia Polytechnic Institute—or Virginia Tech—and when he graduated in 1959, he was kept on at Langley full-time. So, he saw firsthand some of what went on as NACA transitioned to NASA, and he was familiar with the group that had done the initial studies about going to the Moon. He'd also heard of Abe Silverstein, who had become NASA's first director of spaceflight programs. Before any spacecraft or specific details were even hashed out about a potential Moon mission, Silverstein called it Apollo, because he thought the idea of a Greek god riding his chariot across the sun was appropriate to the grand scale of the proposed program. Silverstein's group officially introduced the concept of the Apollo program at a meeting of NASA program planners in November 1959 at the test facility on Wallops Island.

Vaughan had spent many days on Wallops Island himself, working with his fellow engineers in a group called the Pilotless Aircraft Research Division (PARD) and a special research section called the Space Vehicle Group. Through the launches of hundreds of sounding rockets, short-range missiles and eventually larger rockets, they gained a better understanding of the space environment and the hardware required to get there.

Project Echo, NASA's first communications satellite, was a passive spacecraft based on a balloon design created by engineers at NASA's Langley Research Center. Made of Mylar, the satellite measured 100 feet (30 m) in diameter. Once in orbit, residual air inside the balloon expanded, and the balloon began its task of reflecting radio transmissions from one ground station back to another. Echo 1 satellites like this one generated a lot of interest because they could be seen with the naked eye from the ground as they passed overhead. Credit: NASA/Goddard Space Flight Center.

"We tested all sorts of things at Wallops Island in the '50s," Vaughan said. "For example, nobody had a good understanding of the density of the atmosphere as it transitioned into space. That was very much an unknown."

A project designed to provide this type of data utilized inflatable spheres packaged in small containers. They were launched to high altitudes, then released and inflated to a 12-foot (3.6-m)-diameter sphere. Vaughan and his team would study how they performed in the upper atmosphere to measure atmospheric density at different altitudes.

"Their orbit would decay at a fairly rapid rate, but if we could get them high enough and if we could watch them for a couple of days," Vaughan explained, "eventually we were able to document the pressure density environment and provide the data for anyone to utilize."

He then transitioned to working on 100-foot (30-m)-diameter balloons in a project called Echo, one of the first passive-communications satellite experiments.

"Communications were really hard in those days," Vaughan recalled, "so if you were trying to communicate with an aircraft or anything on the ground at a long distance away, you'd go over the cliff pretty quickly where you'd lose the signal because of the curvature of the Earth. The thinking at that time was that if we could launch these big balloons into a geostationary orbit and position about three of those around the globe, you'd have some good communications by bouncing radar signals off them. As an added benefit, the large balloon was easily visible at night for the public to view and we gained a positive response as a result."

However, only one of the giant balloons was launched, because soon, along came better electronics and transistors to replace vacuum tubes, beginning the trend toward miniaturization, more durability and greater computation power. That meant the first active-communication satellites could be conceived of and built. The advances in rocketry meant that these satellites had a chance of being launched into orbit.

And rockets and spacecraft were what excited Vaughan the most. The rocket industry had progressed primarily because of the military's work with intercontinental ballistic missiles (ICBMs): Engineers figured out that instead of lobbing it halfway around the world, if the rocket's trajectory was set just right, it could be sent to Earth orbit. Vaughan witnessed a lot of early failures, but once they got something into space, they had to figure out how to operate a satellite or vehicle in an orbital environment.

Vaughan joined the Space Task Group in the fall of 1961 and, because of his expertise in stabilizing spacecraft, was asked to work on the Gemini and Apollo RCSs. He was hired by Dick Ferguson, the engineering manager for the Energy Systems group, and Vaughan relocated to MSC in early 1962. Since he was single, he was excited to make the move to Houston. But he was singularly focused on his work.

"We needed a better propellant system after we were in orbit," Vaughan said, "better than what Mercury was using. But bipropellants are toxic, very corrosive and were at a low level of technology with respect to all of the required system components. Getting ready for Gemini and then Apollo, we needed better data about the chemical propellants that we needed to use and how to get our reaction control systems to be within the required safety standards and down to the weight restrictions where the Saturn rocket could lift everything."

In 1962, they still had to work out several issues with the propellant, as there were many unknowns and problems they didn't know how to solve or even how to stay away from.

"We had to come up with the concepts for doing all that in space, so first we had to prove we could do it and then number two, we had to provide a workbook for documenting everything," Vaughan said. "But we were blowing up the thrusters, had trouble with the valving. We didn't have a good understanding of how frequently we needed to be able to fire the engines to get the control we needed, especially on the lunar module. And we didn't know the exact impulse we needed."

Specific impulse is a measure of how effectively a rocket uses propellant, and while using puffs of high-pressure air is an easy way to do it, this method doesn't provide much impulse. A single propellant, called a monopropellant, produces a little better power and bipropellants provide the best power.

The fast and repeatable firings of the RCS required not only some unique hardware but also software.

"The avionics for that were just emerging, so we were kind of developing what we needed while we hoped for the software to make it possible," Vaughan said. "The technology was just not there yet."

Vaughan also worked on some analysis with Ferguson on the Apollo escape system, which would provide an exit in case of an emergency at the launchpad. Vaughan's job was to determine the size and scope of it so the details could be included in the statements of work for the contractors to build each system. And keeping the weight down to a manageable level was a continual challenge for all the systems on Apollo.

"When you start designing a new vehicle, guess what? You need info," Vaughan said. "They're gonna ask you, 'How heavy is your system?' 'Don't know.' 'How heavy is that system over there?' 'Don't know.' We had to do a lot of iterations, parametric studies and some pretty informed guestimates to figure out how we were going to build this whole spacecraft and then be able to launch the thing."

Glynn Lunney at his console in the Mission Control Center during an Apollo simulation exercise at the Manned Spacecraft Center in Houston, Texas. Credit: NASA.

GLYNN LUNNEY DROVE HIS 1958 CHEVY convertible down to Texas right after Scott Carpenter's three-orbit Mercury flight on May 24, 1962. Lunney was a NASA flight controller and now had new responsibilities at MSC: to focus on preparing for Apollo. He also needed to prepare for his family's arrival to Houston.

Now, in late June, he was waiting at the Hobby Airport, watching for the plane carrying his wife, Marilyn, and their seventeen-month-old daughter, Jenny. Marilyn was eight months pregnant with baby number two, and as Lunney stood on the tarmac while the plane taxied in, he wondered how Marilyn was going to react to Houston's stifling summer weather. It didn't take long to find out. As she exited the doorway of the plane into the full afternoon sun, Marilyn staggered backward as if the wall of heat and humidity had assaulted her.

"That was her welcoming moment to Houston," Lunney said, "and I'm sure she wondered what she was doing here."

Chris Kraft, Chief of the Flight Operations Division (center), along with Walter Williams, Flight Operations Director (standing) in the Mercury Control Center, Cape Canaveral, Florida, during the Mercury-Atlas 9 (MA-9) mission. Credit: NASA.

Off they drove in the convertible—with no air conditioning—to the house they were renting, also without AC. In the next few weeks before the birth of their son, Lunney would often come home to find Marilyn and Jenny sitting in the bathtub full of cool water, their favorite respite from the heat.

Lunney was only twenty-five, but he had already established himself as a problem solver, getting out in front to tackle some of the issues of operating a human spacecraft. He had begun his career as an aeronautical research engineer in June 1958 with NACA's co-op training program at the Lewis Research Center in Cleveland. He was in George Low's division, and one day a preliminary drawing of the proposed Mercury spacecraft was being passed around the office. Lunney was utterly captivated—in September 1959, he decided to follow Low to Langley and was soon chosen to become part of the initial thirty-five-member Space Task Group.

Within the first year, Lunney became an analyst, working on the technical issues of spacecraft reentry and of how to control where the Mercury spacecraft would land. He soon met fellow Space Task Group member Christopher Kraft, who had been assigned to the Flight Operations Division and tasked with creating flight plans and figuring out how the spacecraft should be operated.

The job was daunting since humans hadn't yet flown in space. But Kraft's early work as an aircraft flight-test engineer became an inspiration. Kraft led the instrumentation teams on the ground during several of the first attempts to break the sound barrier, and he realized that, just like test pilots, astronauts would need a system of communications and support back on Earth during critical phases of the mission. They would also need a ground-based tracking system and instrumentation on board for the telemetry of data from the spacecraft. The concept of a control center to monitor and operate space-flights in real time was born, and Lunney was fascinated when he heard Kraft give a presentation about the idea during a meeting of the Society of Experimental Test Pilots.

"It was a relatively straightforward thing, but no one had done anything like that before," Lunney said. "We were still just a small group of thirty-five, so I got involved, and then we brought on more people to help figure this out."

A small group started putting together the analytics of what hardware was needed in a control center and what information they wanted to get from the spacecraft. How could they tell if the rocket and spacecraft were functioning correctly? How could they make sure the astronaut was still conscious? How would they communicate with a capsule that was on the other side of the planet, moving at 17,500 miles an hour?

Carl Huss worked on the hardware, while Tecwyn Roberts (one of the Canadian Avro engineers), Cliff Charlesworth, John Llewellyn and Jerry Bostick began actively planning the control center and what specific areas of flight they needed to focus on. Over time, they created the key positions in the control center: the Guidance officer, who would monitor onboard navigational systems and onboard guidance computer software; the Flight Dynamics officer, responsible for the flight path of the space vehicle, and the Retrofire officer, who drew up abort plans and was responsible for determining times for reentry. Kraft and his group created several other flight control positions, along with the flight director, who oversaw all the people and systems. One important task was the person who talked directly to the astronauts, called the capsule communicator, or Capcom.

But how to create and perform all these yet-unknown functions? One of Lunney's first assignments was to help with the Control Center Simulation Group, which planned the simulations used to train both flight controllers and astronauts for the experience of human spaceflight.

The Mercury Mission Control Center building at Cape Canaveral, Florida. Credit: NASA.

The goals of Mercury were straightforward: orbit a manned spacecraft around Earth, investigate the pilot's ability to function in space, and recover both the astronaut and the spacecraft safely. To do this meant that those on Earth had to be in nearly constant communication with the spacecraft as it orbited Earth. A group of Space Task Group engineers designed a worldwide communications network, and a monumental effort ensued to set up the Spaceflight Tracking and Data Network, an impressive system of eighteen ground stations scattered around the world on various islands and three continents. Where there wasn't any land to build a station, NASA employed naval ships on three oceans and five aircraft outfitted with instrumentation. A small control team would operate each remote station.

The main flight control room was built inside a modest structure at Cape Canaveral Air Force Station, and the Mercury Control Center (MCC) took shape, with consoles for fourteen flight controllers to direct all aspects of the spacecraft's flight and to monitor the spacecraft's status and the health of the astronaut. The MCC would also coordinate and maintain the flow of communication between all tracking stations and was designed to inform the recovery forces when the capsule would reenter the atmosphere.

The limits of technology at that time meant that while each station could talk to the orbiting astronaut for a few minutes as he flew overhead, the stations on the other side of the world couldn't speak directly to the MCC. Instead, each station used teletype machines to transmit information to the MCC—in very precise language—on the status of the spacecraft and astronaut. At best, they could send twenty words of teletype per minute.

The finest computers of the day were enlisted to help process the data. Since the newfangled transistor-based IBM 7090s were too big and expensive to have at multiple locations, everyone at NASA shared two of these computers, located at the Goddard Space Flight Center in Maryland, while an older vacuum-tube IBM 709 computer was used at the Cape to compute initial launch trajectories.

An interior view of the Mercury Control Center in 1962, prior to the Mercury-Atlas 8 (MA-8) flight. Credit: NASA.

But the telemetry was all analog, so it came down from the spacecraft and was sent to the computers to be processed for things like trajectory and location and then sent to the MCC. But, there was a time delay, so the flight controllers had only a short time—perhaps thirty seconds—to look at the data, understand it and make a decision if the next phase of the mission was go or no-go. Then the MCC would transmit the instructions to the appropriate station so it could be relayed to the spacecraft. The team had to invent a terse, new language where there could be no misunderstandings in the rapid-fire world of flight control.

"Our brains began to be driven by: What do we need to communicate? What's our next opportunity? What do we need to get done in this pass in the way of conversations with the crew? And then, what do we need to do in terms of problem solving?" said Lunney, who became the flight dynamics officer. "And what do we need to do in terms of recommendations or direction to what we were going to do with the flight? This all got tied to these little five-minute circles of communication time with the crew that were spotted around the world, where we had teams of people."

The consoles in the MCC were basic but they got the job done.

"We had little meters for all the telemetry," Lunney said, "and then we had some built-in displays in the console with little numbers, almost like a clock, where we could do things like calculate and display the retrofire times for the various opportunities for bringing the spacecraft back. And when we got in orbit, we would also use the same system to calculate the deorbit times for the various landing points that we had scattered around the world in various oceans that we had recovery forces in."

Dominating the front wall of the MCC was a large world map and two large projector boards. The map used a series of circles to pinpoint tracking stations. To keep continuous track of the Mercury spacecraft, a mini spacecraft model suspended by wires traced its orbit. The projector boards displayed flight measurements plotted by sliding beads. Trend charts displayed the astronaut's condition.

In just a few months NASA raced through nearly a dozen unmanned Mercury flights. Some were successes and some weren't, but they learned with every mission how to be flight controllers. Even the infamous 4-inch (10-cm) flight of the Mercury-Redstone in November 1960 was a learning experience. It was Lunney's first time sitting in the flight control team chair, and suddenly he had to help decide on a completely unexpected course of action. While the fully fueled and armed rocket had been released for flight, it was still sitting on top of the pad, unrestrained by any hold-down device. The parachute draped along the length of the rocket might fill with the Florida coastal breezes and pull the rocket over, which would certainly mean a devastating explosion. The Redstone team in the blockhouse scrambled to decide how to "safe" the rocket. (One idea that was quickly abandoned was using a high-powered rifle to create a hole in the Redstone tank so the fuel would drain out.)

"Up until this event, I had a rather constrained view of what my job as a flight dynamics officer might entail," Lunney said. "All of a sudden, the preparation for effectively operating the MCC took on several more dimensions than I had been imagining. From that day on, my thinking and that of my colleagues embraced the idea that the unexpected could happen and things could get even more complicated from there."

Lunney came to think of the MCC building at the Cape as something close to sacred, "a cathedral of sorts where we went and did what we thought was important work for our country and for humanity." With lessons learned on every unmanned mission, Lunney took his seat at the MCC as the flight dynamics officer for the first two ballistic flights for Alan Shepard and Gus Grissom. Later, he was stationed at Bermuda for John Glenn's orbital flight. When the mission was delayed for two weeks, Lunney also learned about things like rum, card games and scuba diving.

"It was all fun," he said, "basically a crash-course PhD in things we had to invent and figure out. Man, we were busy."

He was asked to move to Houston to become head of the Mission Logic and Computer Hardware section, to define and oversee the computing and display requirements of the Flight Dynamics Division within the new Mission Control Center.

"We had just this handful of people," Lunney said, "and moving to Houston, we were ready to start staffing up. I began hiring these young men right out of college, probably about one a month, putting them to work on figuring out how to monitor the lunar landing. These young men were inventing everything we needed to do."

The new Mission Control would have state-of-the-art mainframe computers from IBM providing the computing backbone for mission operations, allowing the flight controllers to pull up graphs, tables and pictures in seconds and overlay static images (such as maps) with computer-generated images displaying things like spacecraft flight paths. More than two hundred different types of telemetry data would flow into Mission Control, displaying the condition of the astronauts, spacecraft and booster systems.

For interoffice communications, Mission Control created a high-tech, fifty-three-station pneumatic tube system with 2 miles (3.2 km) of tubing and electrically manipulated switches and control valves. This system would automatically guide messages to their final destination.

And finally, teletype equipment that could transmit as many as one hundred words per minute.

It was going to be a new era in spaceflight.

EARLY IN HER CAREER, DOTTIE LEE HAD spent hours on hours punching in numbers on a Friden calculator, computing the trajectories of various rockets. The Arcas, Cajun, Aerobee, Little Joe, Scout—she plotted them all. The noisy Friden contraption was a god-awful green-gray color and took up half her desk at the Langley Research Center. Lee punched the loud, clackety buttons on the electric machine, pulled the lever and hoped with each computation it didn't go into a do-loop—which inevitably it would—meaning she would have to turn it off, unplug it and start over.

"Oh, goodness," she said. "You can't begin to appreciate what it takes to calculate the trajectory of a vehicle on one of those, and you do it for every second for the time in flight. You punched lots of numbers."

Dorothy "Dottie" Lee at her desk, holding a model of the Apollo Command Module. Credit: NASA/Johnson Space Center.

Women scientists gathered in a meeting room at NASA's Langley Research Center in 1959. From left to right, Lucille Coltrane, a computer at Langley Research Center; Jean Clark Keating, an aerospace engineer; Katherine Collie Speegle, a mathematician; Dorothy "Dottie" Lee (standing), mathematician; Ruth I. Whitman, an engineer in the pilotless aircraft division; Emily Stephens Mueller, a computer who worked with the Space Task Group. Credit: NASA.

And even though the calculations spit out by the Friden were helpful, Lee still had to plot everything by hand. The calculations were completed with the aid of a slide rule and results were recorded in logs and plotted on graphs.

But it was exhilarating, and from the beginning Lee had an innate sense that she was in the right place at the right time. Lee started out in June 1948 as a "computer" at Langley, recruited by NACA while at Randolph-Macon Woman's College in Lynchburg, Virginia. She took a position in PARD, and at some point, her job title changed to a math aide in the Computing Section. She ran the calculations for the teams launching the rockets and testing different configurations. The engineers would design a model and fly it, then Lee and the other women in the Computing Section would take the telemetry data and run the aerodynamic calculations for it. Later, Lee worked on calculating some of the initial heat transfer designs for reentering the vehicles—and the numbers showed the future Mercury spacecraft, blunt in shape and built with just the right materials, could plummet down through the atmosphere without burning up.

Over time, Lee worked her way to becoming an aeronautical engineer and later an aeronautical research scientist. That meant that she was designing and testing vehicles, not just running the numbers—she was one of the very few women to do such work. She recalled one test of a five-stage launch vehicle out at the test facility on Wallops. It was so windy, they had to wait all day until the squalls on the barrier island calmed down enough so the rocket's trajectory wouldn't be jeopardized.

"Finally, by dark we were able to launch," she said. "A night launch was exciting because you could see as each stage separated and the next one ignited. But the fifth stage we never saw ignite, and we knew we'd have to wait to get the data telemetered back to us to see what happened."

From Wallops Island they took the ferry back to the mainland. It was midnight, but on the boat Lee and her fellow engineers huddled over their charts, trying to determine why that fifth stage did not ignite. It was a failure, they knew, but the thrill of trying to solve a problem was what Lee relished.

"I can't recall which rocket that was or even the date," she said. "So I don't think it will go down in history as anything except that we learned with each experience. And that's what I did every day of my life."

Lee knew she was lucky that her boss, Max Faget, saw and appreciated her abilities. The fact that Lee was a woman didn't seem to matter to Faget; he valued her skills and how she approached her work. Faget was a research scientist and head of PARD, and Lee shared an office with his secretary, Shirley. When Shirley left for a two-week honeymoon, Lee was asked to fill in on the secretarial duties. While she didn't know how to type, she answered the phone and distributed the mail. And all the while, she was working on a particularly sticky set of triple integral calculations another engineer had asked her to compute. For two weeks she diligently worked the numbers until she solved the problem.

Faget had watched her efforts and at the end of the two weeks said, "Dottie, how would you like to work for me all the time?"

Lee thought he was being funny because she didn't type and she knew Shirley was returning the next week. She flippantly replied, "Oh, sure."

Faget left the room, went downstairs to talk to the division chief and returned, saying she was to start working for him— as an engineer—on Monday.

"They found me a desk, and I was put with some engineers who were beautiful and brilliant and they taught me how to be an engineer," she said. "I learned on the job."

She learned that engineers do their work by testing, calculating and sometimes working backward from the requirements. She discovered she already had fortitude and stick-to-itiveness like the men she worked with and that she, too, liked her facts in black and white—whether they were true or false, good or bad. And she learned that engineers sometimes figure things out in unconventional ways. Lee walked into Faget's office one day in 1958 to find him standing on his desk, dropping paper models of the Mercury spacecraft, checking for the stability of each model. But to her, that was the beauty of the hands-on engineering world she was in. At Langley they lived for data; they could feel it and understand it.

Growing up, Lee somehow knew at ten years of age that the human race was going to the Moon. How she arrived at that notion she's not sure, but she recalls sitting outside one evening looking at the stars and just *knowing*. She was an only child who read voraciously and she particularly enjoyed reading the works of George Gamow and other physicists— authors she noticed other ten-year-olds weren't reading, whether they were girls or boys.

And now here she was helping design a spacecraft that was going to the Moon. But to get to the Moon, she had to move to Texas. So, in the early summer of 1962, she was standing on a muddy lot, overlooking the bayou in Dickinson, Texas, checking in on the construction of the new house she and her husband, John, had designed.

Dr. Maxime A. Faget was the Director of Engineering and Development at the Manned Spacecraft Center for NASA. Credit: NASA.

John, too, was an engineer at Langley and also worked for Faget. They had met in their early days there, were married in 1951 and started their family right away. Langley allowed women to continue working while they were pregnant and raising children, a decision unusual for the times. Dottie loved her work and couldn't imagine not being at Langley. In 1958, with the formation of NASA and the feverish work on Project Mercury, their professional responsibilities increased dramatically. John was one of the first people assigned to the Space Task Group: He was chief of the Mechanical Systems Section for Mercury, overseeing the work done on the parachutes, rockets, pyrotechnics and hydrogen peroxide jets, all of which had to work perfectly for the missions to succeed. Then he was the lead engineer on a study that mapped out the specifics of how NASA could go to the Moon with Project Apollo. John was also on the management committee to choose the contractor responsible for building the Apollo Command and Service Modules. Meanwhile, Dottie was working on determining the requirements for Apollo's heat shield and other systems. Although they worked in different areas, Dottie and John were both committed to Apollo, so they were moving to Houston with their two daughters, Laurie and Dottie Mae.

The first time they saw this plot of land in Dickinson was on a beautiful December day. NASA had flown a group of those who were moving to Houston down from Langley to provide them an opportunity to look for housing. When they left Virginia, it was snowing, but the weather in Houston was warm and gorgeous. John said, "Gosh, what a wonderful place!" But Dottie wasn't so sure. She grew up in Louisiana and knew what the sultry summers could be like down south. She was also having second thoughts about leaving Langley.

The home search was frustrating because the area around the new MSC was almost entirely undeveloped. The couple decided to look for a lot where they could build a house but couldn't find any they liked. The second day in Houston, late in the afternoon, they had given up and gone back to their motel room. Dottie was upset, crying, not wanting to make the move.

There was a knock on the door—it was Faget. He exclaimed excitedly, "I've found it! I've found it, a place to build our homes—let's go!" The three of them drove over to the Kellner Division in Dickinson, where there were two lots available. They were beautiful—overlooking the bayou with big oak trees covered in Spanish moss.

"I'll take this lot, and you take that lot," Faget said enthusiastically. "Now, Johnny, I have to go back to Virginia tomorrow. I want you to buy these lots for both of us first thing in the morning." They got lots.

Back in Virginia, every day over lunch Dottie took out a sheet of graph paper and worked on their new home's floor plan. Now the house was taking shape, on schedule to be done when Dottie and the girls would move in August, in time for the start of school. John was already living in Houston, staying at the barracks at Ellington Field. NASA provided a shuttle flying between Houston and Langley every weekend. Every other weekend, John would fly to Virginia, and on the alternate weekends, Dottie would fly to Texas. But there were times John was so busy, he didn't always get back to Virginia. He was working ten to twelve hours a day, sometimes six or seven days a week. For John, the work was exhilarating. But during one trip to see the family, Dottie told him, "Your daughter Laurie said that she wants to get a new daddy." John took Laurie in his arms and asked her why. Laurie said, "I want one that stays home sometimes."

That was a wake-up call, and he promised the girls that living in Texas was going to be an adventure. But he and Dottie would continue to be busy, working hard to get to the Moon.

EVERYONE AT MSC WAS BUSY, LITERALLY

doing three things at once. In 1962 NASA was flying the Mercury spacecraft, in the process of building the Gemini spacecraft and designing the Apollo spacecraft. NASA had already signed the contract in November 1961 with North American Aviation in Downey, California, for the Command and Service Modules for the Apollo spacecraft. They were also in the process of choosing contractors for the other components and systems.

Gilruth and his staff had to hire enough people to run the three space programs, build the MSC with all its laboratories and test facilities and organize the entire organization with all the directorates and various divisions. Gilruth made Faget head of the Engineering and Development Directorate, responsible for the design, development and testing of the spacecraft and its hardware. Dr. Christopher Kraft was made head of the Flight Operations Directorate, which was responsible for building the Mission Control Center and flying the missions. Astronaut Deke Slayton, who had been grounded because of a heart condition, was appointed head of the Astronaut Office. Guy Thibodaux was the head of the Auxiliary Propulsion Division, while Ralph Sawyer took the helm of the Electronics Division. Joe Shea was sent down from NASA Headquarters to be the program manager for the Apollo Spacecraft Program Office.

There was so much to do in a short period of time, with major decisions to be made at every turn. But in the summer of 1962, one decision loomed above them all: How was NASA actually going to get to the Moon and land there?

In the beginning—even with so little spaceflight experience and so many unknowns—the method to accomplish the goal of getting to the Moon seemed like an easy and logical choice. But soon, heated differences of opinion arose. Friendly scientific discussions became a rivalrous and acrimonious debate after Kennedy had directed NASA to land on the Moon in less than a decade.

It turned out there were going to be three options to choose from. The first—and the early favorite—was called Direct Ascent. This was a vision of spaceflight directly from Jules Verne, H. G. Wells or the science fiction from the 1940s and 1950s: A massive spacecraft launches directly to the Moon, lands "fins first" on the lunar surface using retro-rockets, the astronauts climb down a ladder and do their explorations, then climb up and blast off again. The mission would be completely self-contained in one giant spaceship. At first blush, this method seemed simple in comparison to trying to rendezvous and dock in space, a feat not yet attempted. This fiction-inspired version of the Moon landing seemed to be stubbornly stuck in the heads of almost every decision maker at NASA, and von Braun had already been designing an incredibly enormous booster called Nova that might possibly make it feasible. Direct Ascent was the brute-force method of getting to the Moon.

The second option was called Earth-Orbit Rendezvous (EOR), which entailed two spacecraft launching separately on two different Saturn rockets. The two ships would rendezvous and dock in Earth orbit to create one large vehicle; the ship would then either be fueled in orbit or hook up with a booster rocket, then leave for and land on the Moon in the same manner as envisioned by Direct Ascent. This method took less brute force, as it had the advantage of using the Saturn 1 rocket, which had already had a successful test flight in October 1961. Additionally, the same EOR concept could be used to build space stations or other large space vehicles on Earth—which were among the early suggestions from the study groups at Langley for determining America's future in space. And of course, von Braun had also proposed large, elaborate space stations in Earth orbit. But early on, the concept of rendezvous seemed mysterious and impossible.

"There was a reluctance to believe that the rendezvous maneuver was an easy thing," said Clinton Brown, who headed up a small research group at Langley, the Lunar Mission Steering Group on Trajectories and Guidance. "In fact, to a layman, if you were to explain what you had to do to perform a rendezvous in space, he would say that sounds so difficult we'll never be able to do it this century."

But Brown became one of the instigators for a third option, a dark horse for landing on the Moon called Lunar Orbit Rendezvous (LOR). LOR's basic premise was to launch an assembly of three small spacecraft into Earth orbit with a single powerful rocket—the proposed Saturn V could likely handle the load. The three ships were a command ship, a lunar lander and a supply vessel. The trio would head for the Moon together and, once in lunar orbit, the lander would detach, visit the surface, then return to lunar orbit and rendezvous with the other spacecraft before returning to Earth. Early on, anyone who considered the concept of a rendezvous dismissed it out of hand. But a rendezvous around the Moon? If this tricky maneuver failed, astronauts would be dead, stranded in lunar orbit.

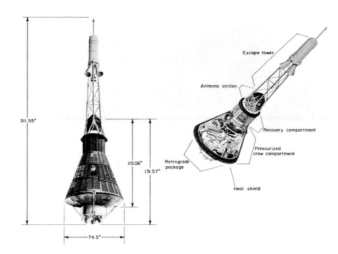

Mercury spacecraft with measurements and cutaway view. Credit: NASA.

Brown had studied the 1920s writings of German physicist Hermann Oberth and other early space theorists. Brown concluded that just building a space station after Mercury wouldn't be considered a significant advancement in the United States' activity in space. So he started a group with fellow engineers William Michael Jr., John Bird, Ralph Stone and Max Kurbju who studied not only a circumlunar trip around the Moon but also a rendezvous to facilitate a lunar landing. Then, during an offhand discussion with Tom Dolan, a representative from one of Langley's aircraft contractors, the Chance Vought Aircraft Corporation, Brown found a compatriot for early ideas about LOR. Brown and Dolan were actually studying two different things—the mechanics of a trip to the Moon and the procedures for a rendezvous in space. During subsequent conversations, noodling sessions, shared scratch-paper sketches and then official meetings, they put their two analyses together to create the idea of LOR for a human lunar mission. But the idea couldn't seem to gain traction with the Direct Ascent and EOR crowds.

Then along came John Houbolt, another engineer at Langley who was working on understanding rendezvous in Earth orbit. The story goes that Houbolt attended a presentation on LOR by Clinton Brown's group sometime in 1960 and it was like a bolt of lightning hit him. He was an instant convert.

Engineer John C. Houbolt explains the Lunar Orbit Rendezvous concept at the Langely Research Center. Credit: NASA.

The more Houbolt studied LOR, the more he was convinced it was the only way NASA could get to the Moon. LOR required less fuel, only half the payload and not as much new technology compared to the other methods. It would take just one launch from Earth, as opposed to EOR's two or more. Only a small, lightweight Lunar Module would land on the Moon instead of a massive rocket ship. And because the lander could be discarded after use, it wouldn't have to be built to withstand the fiery return through Earth's atmosphere. Instead, it could be designed for maneuvering in the lunar environment, which had no atmosphere and only one-sixth the gravity of Earth.

Houbolt became a fervent champion of LOR. Charlie Donlan, then one of the leaders at Langley, equated Houbolt to John the Baptist, traveling the wilderness of ad hoc committee meetings and scientific conferences, passionately preaching the LOR gospel to anyone who would listen. He was trying to save the souls of everyone at NASA because he was sure the only way to attain the salvation of landing on the Moon was through LOR.

Houbolt first proposed the idea of LOR to a larger group of engineers during a Space Task Group meeting at Langley in early 1961. He was told he didn't know what he was talking about. In a later meeting, Houbolt was told the numbers he presented were lies.

The reluctance about rendezvous in general continued. At one meeting in June 1961, Abe Silverstein, who was now director of the Office of Space Flight Programs at NASA Headquarters, emphatically said there was no way rendezvous should be part of getting to the Moon.

"Look, fellas, I want you to understand something," he said, starting the meeting. "I've been right most of my life about things, and if you guys are going to talk about rendezvous, any kind of rendezvous, as a way of going to the Moon, forget it. I've heard all those schemes, and I don't want to hear any more of them, because we're not going to the Moon using any of those schemes."

But that didn't stop Houbolt. He kept evangelizing for LOR, and when he was summarily dismissed, he refined his presentations, making them more precise and emphatic.

In the meantime, as engineers continued to study and work out the rendezvous concept, it became apparent that it wasn't going to be as impossible as it first seemed. The realization hit that any missions subsequent to Mercury were going to require it, so they better damn well figure it out. Little by little, the prevailing perceptions against rendezvous began to change. EOR now moved ahead and became the leading candidate for landing on the Moon, even though there wasn't a clear concept of how to set down a large vehicle on the then-unknown lunar surface and then boost it from the Moon's surface back to Earth.

After even further study, some engineers admitted that a rendezvous around the Moon might not be as dangerous as everyone first thought. Early comments about LOR basically being a death sentence for a crew were softened when John Bird from the LOR team refined his comeback: "No, it's like having a big ship moored in the harbor while a little rowboat leaves it, goes ashore and comes back again."

When NASA's new associate administrator, Robert Seamans, came to visit Langley in September 1960, Houbolt intercepted him in a hallway to discuss LOR. Seamans had begun to understand how future missions would depend on the ability to rendezvous, and he invited Houbolt to present his ideas at NASA Headquarters in Washington.

His new report, *Manned Lunar Landing through Use of Lunar Orbit Rendezvous*, outlined the benefits and methodology of a Lunar Orbit Rendezvous mission. But again, his audience did not initially perceive the idea as a realistic and credible way to get to the Moon.

Houbolt continued his crusade throughout 1961 and, feeling desperate that NASA was about to make a disastrous decision, he a wrote a letter directly to Seamans, passing over all official channels. In the nine-page letter, Houbolt referred to himself "somewhat as a voice in the wilderness," then proceeded to outline the challenges facing Apollo and highlight the simplicity, the cost-effectiveness and (most importantly) the timeliness of a lunar rendezvous mission.

"The greatest objection that has been raised about our lunar rendezvous plan is that it does not conform to the 'ground rules,'" Houbolt wrote. "This to me is nonsense; the important question is[:] Do we want to get to the Moon or not? Why is Nova, with its ponderous size simply just accepted, and why is a much less grandiose scheme involving rendezvous ostracized or put on the defensive?"

Seamans passed the letter along to others at NASA Headquarters, and a copy of it landed on the desk of Joe Shea. Shea's boss, Brainerd Holmes, directed Shea to figure it out once and for all and Shea decided he just needed to go where the data took him.

In the months that followed, more than seven hundred scientists, engineers and researchers in government, industry and academia spent more than a million person-hours studying the various concepts. Decision makers dedicated meetings and conferences to considering the various mission proposals. The choice would have an impact on how the spacecraft and systems would be built and if NASA was going to meet the challenge of getting to the Moon by the end of the decade, the decision needed to be made soon.

NASA had been built on engineering. Since many of the organization's top people were engineers, they couldn't ignore what the numbers now proved. When Direct Ascent was finally examined realistically, it didn't take much more than back-of-the-envelope calculations to determine the development of an enormous rocket would be prohibitively expensive. Plus, the Nova rocket was projected to be so powerful that it could not launch from Cape Canaveral (one engineer only half jokingly said the rocket would sink Florida's Merritt Island). Additionally, the rocket would have to carry

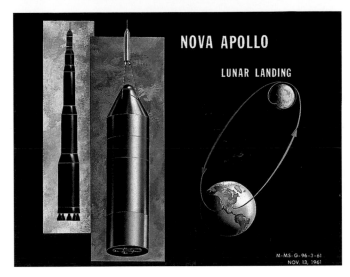

This artist's concept illustrates the Nova launch vehicle concept, which, from 1960 to 1962, the Marshall Space Flight Center considered as the best means to achieve a human lunar landing with a direct flight to the Moon. Although the program was canceled after NASA planners selected the Lunar Orbit Rendezvous mode, the proposed F-1 engine would eventually be used in the Apollo Program to propel the first stage of the Saturn V launch vehicle. Credit: NASA/Marshall Space Flight Center (MSFC).

a massive amount of fuel to be able to land and then lift off from the Moon. And based on the limited success of the United States' early small rockets, a Direct Ascent vehicle would probably take decades to design and build; it would not qualify for Kennedy's challenge.

As the calendar turned to 1962, in many people's minds, the decision was coming down to a battle between the strong cultures of the two NASA centers most involved: MSFC favored EOR since it meant constructing more rockets. MSC came to favor LOR because it meant developing lots of spacecraft. It was going to come down to which strategy was more feasible: landing something big on the Moon or a risky lunar rendezvous.

On June 7, 1962, the Lunar Mode Decision Conference was held at MSFC. Representatives from Marshall spent four hours in the morning presenting their argument for EOR, and in the afternoon the Houston contingent spent another four hours stating their case for LOR. As the day wore on, it became apparent to most everyone which approach was more viable. In the end, von Braun stood up and, to the surprise of his Marshall colleagues, expressed support for LOR:

Catherine Osgood, NASA math aide and aerospace engineer. Credit: NASA.

We at the Marshall Space Flight Center readily admit that when first exposed to the proposal of the Lunar Orbit Rendezvous Mode we were a bit skeptical—particularly of the aspect of having the astronauts execute a complicated rendezvous maneuver at a distance of 240,000 miles from the earth where any rescue possibility appeared remote. In the meantime, however, we have spent a great deal of time and effort studying the [three] modes, and we have come to the conclusion that this particular disadvantage is far outweighed by [its] advantages.

We understand that the Manned Spacecraft Center was also quite skeptical at first when John Houbolt advanced the proposal of the Lunar Orbit Rendezvous Mode, and that it took them quite a while to substantiate the feasibility of the method and fully endorse it.

Against this background it can, therefore, be concluded that the issue of "invented here" versus "not invented here" does not apply to either the Manned Spacecraft Center or the Marshall Space Flight Center; that both Centers have actually embraced a scheme suggested by a third source . . . I consider it fortunate indeed for the Manned Lunar Landing Program that both Centers, after much soul searching, have come to the identical solutions.

Although there would be a few holdouts expressing their dissenting opinions well into the next few years, the final decision was that LOR provided the best solution for landing on the Moon before the decade was out. On June 22, 1962, the Manned Space Flight Management Council announced in favor of the LOR approach and the final paperwork went to NASA administrator James Webb for his signature.

WITH THE LOR DECISION NOW FIRMLY

made, Ken Young and his compatriots in the Rendezvous Analysis Branch had an enormous task before them. They were starting with a blank sheet of paper, breaking new technical ground.

"In the beginning, we really didn't know what we were doing," Young said. "We knew how to work the equations of motion, and plan and maneuver the right orbits, but as far as rendezvous goes, we had to find a solution for relative motion."

The questions of relative motion dealt with how two spacecraft moved in regard to one another and not how they were moving around Earth. But the fact that both vehicles would be moving at incredible speeds—about 17,500 miles per hour in low Earth orbit—created complications. In the unusual conditions of orbital flight, navigation is entirely different from navigation on Earth, and the differences are hard to reconcile with the everyday human experience; therefore, the team couldn't use intuition to figure this out.

The intricacies of orbital mechanics meant that after a spacecraft reached orbit, the astronauts couldn't just "floor it" and continuously fire their thrusters in order to meet up with another spacecraft. Doing so would quickly deplete the fuel, and they would likely never reach their target.

Orbital velocities are somewhat counterintuitive, as different orbital altitudes correlate to a certain orbital velocity: The lower the orbit, the higher the orbital velocity; and the higher the orbit, the slower the orbital velocity. If a spacecraft fires its thrusters—seemingly to speed up—and achieves a higher orbit, it ends up slowing down because higher orbits have a lower orbital velocity. And just the opposite—if the spacecraft decreases speed, it reaches a lower orbit, where it ends up going faster, relatively speaking.

The Rendezvous Analysis Branch had to account for Newton's laws of motion and his law of gravity. The branch also had to understand the Hohmann transfer, a maneuver that moves a spacecraft from one orbit to another in the most efficient manner. Other specialized maneuvers, while not as efficient in fuel usage, had to be developed to put a spacecraft into the preferred position relative to another vehicle at a specific time or orbital position.

The key to orbital rendezvous maneuvers is timing the thruster firings so that the two spacecraft arrive at the same point in an orbit at about the same time. Deciding when to fire the thrusters takes quick calculations—somewhat similar to how a quarterback leads a receiver in a football game—to account for how fast and how far the spacecraft needs to go to make the rendezvous.

Equally as important as the orbital maneuvers is the precise timing of the launch of the second vehicle (called the chaser) since it must match the orbital plane of the first spacecraft (called the target vehicle) at the desired time of rendezvous. Thus, many other factors involving the launch window, the launch site, the range safety constraints, the booster lifting and steering capabilities and abort constraints must be researched and analyzed.

And working out the numbers would be very different whether you were computing for Earth orbit versus lunar orbit since the orbital periods and gravitational and environmental conditions were very different. For example, it takes ninety minutes to orbit the Earth and two hours to orbit the Moon.

"You start out with your equations of motion and how one vehicle moves and how the other one moves in relation to it," said Catherine Osgood, who joined the group shortly after Young. "Then you just keep working on it until you've figured out how to lift off at the right time to rendezvous with a vehicle that's already up there. You learned to keep your pencil sharp."

Osgood worked at Langley as a math aide. Shortly before moving to Houston, she applied to become an aerospace engineer, based on all her work experience. She didn't really expect to get her classification and job title changed since she had never taken any engineering classes.

"But then I actually got it," she said. "I was really surprised, but there was no rough transition. It seemed if someone were capable of doing something, they'd be asked to do it, regardless of what their positions were. So I just sort of slid into it."

Osgood's husband, Donald, had worked for the Mercury Tracking and Ground Instrumentation Unit, setting up the communications network around the world for the Mercury flights. For Gemini and Apollo, he would be working with the instrumentation in the new Mission Control Center. The move to Houston meant new jobs, new challenges and living in a part of the country they'd never even visited. After living among the native Tidewater Virginians—who had the reputation of being a somewhat insular people—the Osgoods hoped that Houston might be a little more cosmopolitan.

But it meant moving their three children and finding new arrangements for their care. Both Catherine and Donald traveled for their jobs at various times, and their regular work schedules demanded long hours. They were eventually able to find a live-in nanny, and they also enlisted the help

Bill Tindall, standing and Chris Kraft in Mission Control in Houston in 1973. Credit: NASA.

of the grandmas. Catherine's mother came and stayed with the family in the summertime and Donald's mother came in the winter. All the complications of moving, finding a place to live and getting the kids situated seemed to seamlessly work themselves out so Catherine could concentrate on her work with the rendezvous analysis team.

And this combination of math, physics and geometry took a tremendous amount of concentration. At this point, rendezvous was theoretical, because no one had actually performed any type of maneuver like this in space. In 1962, spaceflight was still in its infancy and humans had just recently made it to orbit. Nobody knew what the unknowns were, but the group at MSC knew rendezvous was going to require some new technologies and procedures. The Gemini spacecraft needed to be able to maneuver with incredible precision and communicate with a better ground tracking system. An in-flight radar system for the spacecraft would need to be developed, and a small computer was going to be essential in calculating complicated rendezvous maneuvers and performing the precise thruster firings. Apollo would need an entirely new set of calculations, which meant the team needed to add all the elements of a new and different spacecraft flying in the lunar environment.

So the rendezvous analysis team worked their slide rules, argued regularly and then went back to working the numbers. While a few engineers at other NASA research centers also worked on resolving the theoretical complexities of rendezvous, the group at MSC had one person no one else had: Bill Tindall.

Tindall graduated from Brown University in 1948 after a stint in the Navy, uncertain what he wanted to do with his mechanical engineering degree. Either by luck or a stroke of kismet, he saw a brochure about the Langley Research Center; it immediately struck a chord. In his first job there, he figured out problems with aircraft wind tunnel instrumentation—but when *Sputnik 1* launched, he felt pulled to the space side of Langley. He moved on to the Project Echo communications satellite, where his bent for dealing with sticky problems soon caught the attention of the Space Task Group. The way he could instinctively unravel a wide variety of issues made it seem like spaceflight was part of his soul.

Truly a man for all seasons, Tindall tooted the trumpet, acted in local theater, played a fiercely competitive game of tennis and unequivocally loved sports cars. Some called him Wild Bill; others thought he was the friendliest, funniest guy they'd ever met. But everyone considered him a genius.

"That guy was so dynamic, it was unbelievable," Ken Young said. "He was an amazing first boss because he was a can-do, get-things-done, get-to-the-point kind of person. We had some great arguments trying to make plans for Gemini and get it done. But he was full of insights to help us younger guys with the subject of orbital mechanics."

Tindall seemed to have a knack for finding the right people with the right talents and then pushing them to do their absolute best. If anyone in the rendezvous group made a comment or decision without citing every single fact or without thinking things through, Tindall could chew them out and take them under his wing at the same time. He had the ability to see the broad view of the entire mission in order to solve the specific problem of rendezvous.

"Bill was brilliant, enthusiastic and energetic and he completely engaged all viewpoints," said Glynn Lunney. "He could tell people to sit down and be quiet when he needed them to, but he also had a way of coaxing out details when people had trouble explaining something. We weren't always articulate, and sometimes we got pretty emotional about our viewpoints."

Early on at Langley, Tindall was a friend and mentor for Lunney, and Lunney watched as his friend became involved in almost every aspect of Mercury. In addition to working on mission planning and trajectories for the first Mercury flights, Tindall's insights dramatically simplified the operations concept for the MCC in Florida and the remote stations around the world. His ideas promoted improvements in the worldwide communications network, and he was now providing valuable inputs for the new Mission Control Center being designed at MSC. In fact, it was Tindall who suggested to Christopher Kraft that Mission Control should be in Houston, where everyone was going to be living, instead of keeping it in Florida. While focused on making everything as streamlined as possible, Tindall understood and valued the importance of home and family.

It was the Tindall family—Bill, Jane and their four children—who first welcomed the young Lunney family to Virginia, and the Tindalls helped the Lunneys with their new home in Houston. Their children played together, and the families began a lifelong connection and commitment to each other.

Tindall provided his expertise in a variety of areas as NASA was planning for Gemini and Apollo, and somehow always got plugged in to the point position for the most challenging subjects—which was why he was handed the job coordinating the Rendezvous Analysis Branch. His approach was to systematically start through all the mission phases and then narrow things down for the specific missions, exercising different rendezvous techniques. His process reduced complexities down to easy-to-understand building blocks.

The rendezvous team started having weekly meetings, during which Tindall asked each person to write down what they'd learned that week. Tindall himself would make meticulous notes on various sheets of paper and then dictate them to his secretary, Patsy Sauer, who would have everything compiled, typed out and distributed by the next day. Eventually, all the input helped create the team's official handbook. Tindall would review the "Rendezvous Notes," making modifications as necessary. It was all a work in progress.

"No, no, no, that's not quite right, where's the original?" Tindall stomped around the office, looking in the trash for his original handwritten notes. "Now, this is what I want sent out."

The frequency of the meetings increased and then began encompassing other aspects of the missions as Tindall saw that rendezvous was just one piece of the entire puzzle. The meetings turned into official events called the Trajectory and Orbit Panel, then evolved into what Tindall called Data Priority, since one of the early issues was which sources of navigation data and other mission-critical information should be used for the particular phases of the missions. And then they had to come up with a plan for how to make all those choices in real time.

Group shot of the nucleus of the 1962 Flight Operations Division for the Mercury program. Image taken at the Houston Petroleum Center in Houston, Texas, prior to their move to the Manned Spacecraft Center (MSC). The women are (left to right) Doris Folkes, Cathy Osgood, Shirley Hunt and Mary Shep Burton. The men are (left to right) Dick Koos, Paul Brumberg, John O'Loughlin, Emil Schiesser, Jim Dalby, Morris Jenkins, Carl Huss, John Mayer, Bill Tindall, Hal Beck, Charlie Allen, Ted Skopinski, Jack Hartung, Glynn Lunney, John Shoosmith, Bill Reini, Lyn Sunseith, Jerry Engel, Harold Miller and Clay Hicks. Credit: NASA.

That's when the flight control team started getting involved in these discussions, involving other mission planners. Then the meetings quickly evolved into a more comprehensive process that included several other systems, such as the flight software team and the experts from the Flight Crew Operations Division who were creating flight checklists and flight plans.

"Then, most significantly, the flight crews enthusiastically got engaged," said Lunney, "and it was a forum in which we systematically talked through and vehemently argued about every step and decision in the process, precisely defining all the flight techniques necessary to use the best of the spacecraft capabilities to accomplish backup launch, guidance, rendezvous, docking, docked propulsion burns, deorbit and entry."

Tindall would orchestrate the discussion and arguments, encouraging everyone's input and often sketching out the details on a blackboard. If the debate turned spirited or heated, Tindall would declare that whoever held the chalk at the end of the meeting was the winner.

"This one talented individual started coordinating other talented people to get the job done," Catherine Osgood said, "and Bill would just dive right into any problem he found, no matter where it was. It was really fascinating to be in that sort of environment."

It was in this atmosphere of resolving the most intricate details for flight operations that the Gemini and Apollo missions started taking their specific shapes.

The original seven Mercury astronauts, each wearing new cowboy hats and a badge in the shape of a star, are pictured onstage at the Sam Houston Coliseum. A large crowd was on hand to welcome them to Houston, Texas. Left to right are astronauts M. Scott Carpenter, L. Gordon Cooper Jr., John H. Glenn Jr., Virgil I. Grissom, Walter M. Schirra Jr., Alan B. Shepard Jr. and Donald K. Slayton. Credit: NASA.

WITH THE MSC CONTINGENT GROWING in numbers, Houston decided to crown itself as "Space City, USA." The chamber of commerce coordinated a huge Fourth of July event to welcome all the newcomers to their community. The day started with a thirty-six-car parade: A police escort led the astronauts and their families, along with several hundred NASA engineers, scientists and other "Space Age celebrities," as the chamber of commerce promoted them, including several prominent politicians who helped bring the MSC to town. Rounding out the parade was a genuine Mercury spacecraft mounted on a special trailer. Lines of convertibles curved through downtown Houston, with the distinguished passengers waving at thousands of Houstonians who came out to cheer.

The motorcade arrived at the Sam Houston Coliseum. Inside the cavernous exhibit hall came speeches from politicians. (Senator John Tower proclaimed, "We rejoice in your presence here; we like you and hope you'll like us.") There was a mass barbecue and the local Texans were decked out in cowboy hats, boots and big belt buckles, but they handed out hats and sheriff's badges to the newcomers. Strains of "Deep in the Heart of Texas" and "The Eyes of Texas" blared from local high school bands. A variety of entertainment included a family-friendly version of a fan dance by renowned stripper Sally Rand.

More speeches came from MSC director Gilruth, deputy director Walt Williams and his assistant Paul Purser. Shorty Powers, the man known as the "Voice of Mercury Control," introduced the seven astronauts to an adoring crowd of about five thousand. *Houston Magazine* touted the event as the "most meaningful Fourth of July parade that Houston has had in recent times."

"They were genuinely happy to have us here in Houston," said Glynn Lunney, "and they made sure that we had all the beer and barbecue that we could handle. This was an amazing change from the locals in Virginia. The people of Houston seemed to love everything about the idea of space and the fact that we were moving into their community. We could not have imagined a more friendly welcome."

Meanwhile, construction of several buildings at the new MSC site started in earnest during the summer of 1962, and Ken Young would drive by the site regularly to check out how things were progressing. It was fun and exciting to see this place come together, just sort of popping up out of the prairie. Two different construction companies were hired to turn the cow pasture into a college campus–like setting by first bringing in the needed utilities and then creating a maze of new streets. The first buildings under construction were Building 1, the headquarters of MSC, which housed the senior management; Building 2 was the public affairs office, with media briefing rooms and other facilities; Building 3 housed the all-important cafeteria and Building 4 would be home to Mission Operations offices for the flight directors and the Flight Crew Operations Division, which included the Astronaut Office.

Mainly due to the persistence of Bill Tindall, NASA transferred its solid-state IBM 7094 computer from Langley to temporary facilities at the University of Houston so that its computing power could be put to work analyzing the new systems being designed and built in the temporary facilities scattered across southeast Houston.

Earlier in the year, Tindall had started a memo-writing campaign to consolidate NASA's computing systems in Houston. He argued that since the MSC would soon be home to the flight control team and because Gemini would be a more complex program, keeping the computers (and programmers) at Goddard and the flight controllers at Cape Canaveral created serious problems for communications and efficiency. He also made recommendations for the new computer center that was going to be built in Houston, called the Real Time Computer Complex (RTCC). Eighteen companies were bidding on building the facility and providing the computers for the RTCC, and in the fall of 1962, IBM was awarded the contract.

THERE'S AN APOCRYPHAL TALE FROM

September 12, 1962, the day that President John F. Kennedy visited the MSC in Houston. The story goes that during the visit, Kennedy noticed a janitor carrying a broom. Interrupting his tour, JFK walked over to the man and said, "Hi, I'm Jack Kennedy. What are you doing?"

"Well, Mr. President," the janitor responded, "I'm helping put a man on the moon."

This anecdote has come to exemplify that no matter how large or small the role, everyone can contribute. And it also illustrates that NASA had come to feel that every job was vital to the effort, and every employee at MSC was dedicated to the mission of Apollo.

But what really happened on September 12, 1962, is that a young spacesuit engineer named Joe Kosmo saved Wernher von Braun's life. But first, Kosmo had to give the president some spare change.

Kennedy was on a whirlwind two-day tour of the country's space installations. He'd brought an entourage to see the launchpads and other facilities at Cape Canaveral in Florida, and he'd toured MSFC in Alabama where the big Saturn V rocket was being designed and built. Now he was in Houston to visit some of the temporary MSC facilities and get an update on how work was progressing for Apollo.

Kennedy, Vice President Johnson and about thirty other dignitaries—NASA officials, politicians and military advisers—sat in on an hour-long classified briefing. Knowing of Kennedy's back problems, Gilruth had arranged for a rocking chair for the president. When JFK entered the room, he smiled when he saw the chair, took off his suit coat and threw it to his assistant. Then he rolled up his shirt sleeves and sat down comfortably.

Kosmo was there, wearing a fully pressurized spacesuit, waiting his turn during the show-and-tell-type demonstrations of the top-secret strides being made in regard to space. Even in the spacesuit, Kosmo felt out of place. He looked around the room at all the famous people and felt equivalent to the guy with a shovel who comes behind the elephants in a parade.

President John F. Kennedy delivering a speech at Rice University Stadium in Houston, Texas, on September 12, 1962. Credit: NASA.

Then came his turn to stand up in front of the group and show off some of the advances in this new version of the Apollo prototype suit, while his division chief, Dick Johnston, gave an overview of the spacesuit and the life-support system.

All of a sudden, Johnston reached in his pocket, threw a quarter on the floor and said, "Watch this, Mr. President, as Joe can now pick up the quarter."

Kosmo's eyes shot open. What the hell was Johnston doing? They hadn't practiced this. They had never even tried anything like this before, and now—in front of all these people—he hoped he damn well could bend over. Kosmo is relatively tall, and the suit's waist joint wasn't the greatest thing the spacesuit team had ever designed. He shimmied over to one side and gradually leaned over. After some effort, somehow he was able to reach the floor and, even with the bulky gloves, pick up the quarter. Not sure what to do next, Kosmo handed the quarter to a delighted President Kennedy. The group applauded. Kosmo breathed a sigh of relief.

Later, there was a larger event for the press and MSC employees held in MSC's current Spacecraft Research Division facility, which was the former Rich Fan Manufacturing Building. When first acquired by NASA, this had been a rather drab industrial plant, but NASA had painted it white and put a large "Manned Spacecraft Center, Site 3" sign and NASA logo out front. Two stories of offices lined one side of the building with a complex maze of halls and partitions, and there was a large, open bay where fan-making equipment once stood. When Kennedy's visit was announced, NASA got busy painting corridors, installing new flooring and telling workers to tidy their workspaces.

Left: Dr. Robert R. Gilruth (left), director of NASA's Manned Spacecraft Center, and President John F. Kennedy look at a small model of the Apollo Command Module during Kennedy's visit to Houston in 1962. Credit: NASA.

The big bay was now filled with mock-ups of a Gemini spacecraft, two different Apollo Command Module designs and a Lunar Module (reporters called it the "Lunar Bug" because of its odd shape), as well as the actual *Aurora 7* Mercury capsule flown by Scott Carpenter. A vibration table and numerous other pieces of test equipment were on display. Several of the astronauts—Grissom, Carpenter, Shepard, Glenn and Slayton—were present. There was NASA administrator James Webb, MSC director Gilruth, von Braun and other NASA leaders; a few local politicians; Texas governor Price Daniel; Kennedy and Johnson and several military officials including Air Force general Curtis LeMay.

Kennedy sat inside one of the Apollo Command Modules with Slayton, then a queue of people waited to climb up a set of stairs to look inside the Lunar Module. Kosmo nervously noted that von Braun was ahead of him in line. After the German engineer had climbed the stairs and peered inside, as he turned to come back down the stairs, he stumbled and fell. Kosmo saw the whole thing and reached out to grab von Braun, who landed on top of him, cushioning the fall. The crowd of people gasped and rushed to check if the two men were okay. As Kosmo helped von Braun to his feet, the stunned rocket engineer straightened and brushed off his suit, looked Kosmo in the eyes and said, "Thank you, young man, I think you just saved my life!"

Kosmo figured he would probably remember this day for the rest of his life.

But this day would become memorable to many Americans for another reason. For those who were present, their most vivid recollection might be how quintessentially hot and humid it was at Rice Stadium, where the president gave a speech. Kennedy himself compared the day's heat to the temperatures experienced by a spacecraft reentering the atmosphere. But what everyone else recalls—and perhaps holds dear—are the president's now-famous words:

> Its [space's] conquest deserves the best of us all and its opportunity for peaceful cooperation may never come again.

> But why, some say, the moon? Why choose this as our goal? And they may well ask: Why climb the highest mountain? Why 35 years ago fly the Atlantic? Why does Rice play Texas? We choose to go to the moon in this decade and do the other things, not because they are easy, but because they are hard—because that goal will serve to organize and measure the best of our energies and skills—because that challenge is one we are willing to accept, one we are unwilling to postpone, and one we intend to win.

After the eighteen-minute speech, the crowd rose in a resounding ovation, then left the stadium galvanized in this historic effort to reach beyond Earth. Kennedy's words propelled the space program to the forefront of American culture and consciousness, and what was once deemed impossible was now considered something that absolutely had to be accomplished. The president had asked the dreamer in all Americans to imagine a new and bold future.

AS THE HEAT OF SUMMER BEGAN TO wane and autumn approached, everyone at MSC began to fall into their new routines of work, family and more work. The rendezvous team made significant headway in their calculations. Dottie and John Lee moved into their new home. Glynn and Marilyn Lunney named their son Glynn Jr. And, after some reorganization at the MSC, Glynn Sr. was named the section head of the Mission Control Center Branch. For the time being, just one other person—Lunney's friend Cliff Charlesworth—shared his office. Norman Chaffee continued his work with Henry Pohl, Chester Vaughan and other young engineers, sometimes getting input from their division chief, Guy Thibodaux.

"Then something would come up, and we'd have to go talk to the program manager or something like that, and I'd go up and there would be this room full of senior folks," Chaffee said. "I'd get to say what the division thought, knowing that my division management stood behind me. I was always given a very polite, thoughtful audience by those people. I was thanked or often was questioned and queried. For a twenty-six-year-old kid being in this situation, it was remarkable."

Chaffee was sure that in no other set of circumstances could he have this type of experience: to become an expert in an area where there were few experts, to be involved in a project of such importance, to be challenged with all he could do and more but also feel he was getting away with the greatest heist in the world and doing exactly what he was meant to do.

Portrait of the first two groups of astronauts. The seven original Mercury astronauts plus new members of the astronaut corps. Seated from left to right are Gordon Cooper, Gus Grissom, Scott Carpenter, Wally Schirra, John Glenn, Alan Shepard and Deke Slayton. Standing from left to right are Edward White, James McDivitt, John Young, Elliot See, Charles Conrad, Frank Borman, Neil Armstrong, Thomas Stafford and James Lovell. Credit: NASA.

NASA KNEW THEY WOULD NEED MORE astronauts to accomplish all the upcoming flights with the new challenges of rendezvous and lunar landing, so in mid-September, a new batch of astronauts was announced: Neil Armstrong, Frank Borman, Pete Conrad, Jim Lovell, Jim McDivitt, Elliot See, Tom Stafford, Ed White and John Young. All had test pilot experience and four had advanced engineering degrees.

The remainder of the year unfolded like a preview of what the rest of the 1960s would hold. On October 4, Wally Schirra's Mercury-Atlas 8 flight launched from Cape Canaveral and returned to Earth after six orbits; in December, the US spacecraft *Mariner 2* passed by Venus, becoming the first spacecraft to transmit data from another planet; and that fall, the cartoon *The Jetsons* debuted on TV. The space age was irrevocably here.

Television became a central part of more American homes, so not only could people see NASA's every rocket launch but viewers watched live—sometimes in horror—as political and racial tensions erupted across the country in riots and demonstrations, some turning deadly. The nightly national news reports also stoked Cold War fears, especially during the thirteen-day Cuban Missile Crisis, a political and military standoff in October that brought the United States and the Soviet Union to the brink of nuclear war. Even though President Kennedy and Soviet Premier Nikita Khrushchev negotiated a peaceful outcome, tensions and suspicions continued between the two countries.

Not as well covered by the national news were fires that broke out in two separate tests simulating the length of an Apollo mission to the Moon. These NASA experiments were conducted to help determine the effect on the astronauts of breathing pure oxygen for fourteen days. Even though the Air Force and Navy participants were injured in these fires, NASA decided they still favored the use of pure oxygen inside spacecraft.

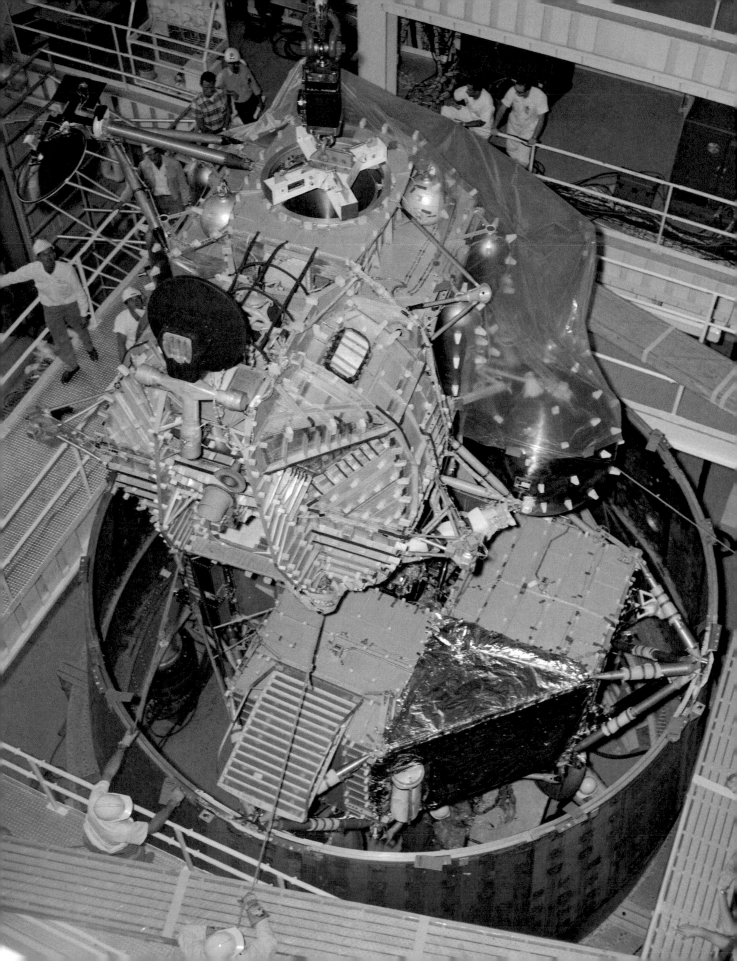

CHAPTER 2

1963

Just go like crazy and get the job done.

—BOB WREN, lead test engineer for the Manned Spacecraft Center's Vibration and Acoustic Test Facility

NAVIGATING A SPACECRAFT TO THE MOON

and back was going to require a computer. It was that simple. But in the early 1960s, putting a computer inside a spacecraft was not at all a simple proposition.

"Computers were huge, room-size things," said Dick Battin, who worked at the Instrumentation Laboratory at the Massachusetts Institute of Technology (MIT) in Cambridge. "The idea of squeezing one inside a spacecraft seemed preposterous."

Battin knew that only a computer could solve the dynamic equations of motion required for accurate navigation of space vehicles operating at high speeds. Because of the complexities and distances of spaceflight, such a computer would need to operate autonomously and in real time. But most of the computers considered real time in the early 1960s were analog—with vacuum tubes that failed frequently and required huge amounts of electricity. Those requirements weren't feasible in a spacecraft, and digital computers of the day were just not fast enough to do real-time computations. In order to meet the national goal of landing humans on the Moon, NASA would need to find a capable navigation system and computer.

MIT's Instrumentation Lab had been working on small computers for military and aerospace use since the mid-1950s. Dr. Charles Stark Draper, a gregarious and larger-than-life engineer and longtime professor of aeronautics at MIT, founded the lab in the 1930s. At first, the Lab allowed Draper's students to get hands-on experience with things like wiring fuel and altitude gauges for airplanes—but over time it became a full-on laboratory, developing instrumentation for aircraft navigation. During World War II, the Lab developed gyroscopic equipment that led to the gunsights used by the Navy's antiaircraft weaponry, which subsequently led to the guidance systems for Cold War intercontinental ballistic missiles (ICBMs).

With *Sputnik 1*'s launch, Doc Draper wanted to get involved with spaceflight too. And with his blessing, a small group at the Lab started working on a secret project, a small little spacecraft they called the *Mars Probe*.

Left: The Apollo Lunar Module being installed in the Vibration and Acoustic Test Facility at MSC in Houston, Texas. Credit: NASA.

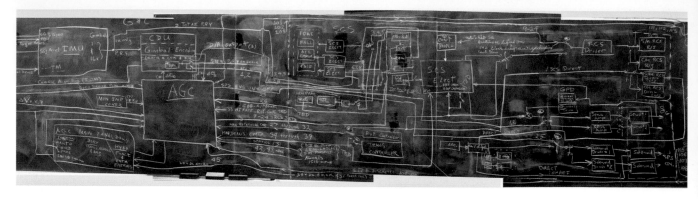

Composite image of a block diagram of the hardware formula for the Apollo Guidance Computer for the Command and Service Modules, drawn on a chalk board at NASA, drawn up by MIT engineers. The diagram depicts how the different systems and components communicate with each other. Signatures of the different participants are on the right side (NASA, MIT and NAA [North American Aviation]). Credit: Draper.

"We were trying to design a spacecraft and the guidance system for a round-trip flight to Mars," said Battin. "It would take a picture of the surface of Mars and bring the picture back to Earth. We suspected that if we were ever going to get into the space business, we ought to do it right away. Rather than sitting there waiting for somebody to ask us to do something, we decided to make a dramatic proposal ourselves."

The group included Milt Trageser, who led the spacecraft design; Hal Laning, who performed preliminary calculations of trajectories to Mars; and Battin, Eldon Hall, Ralph Ragan, David Hoag and a few others, who performed studies on what would be required for guidance and navigation. The *Mars Probe* team realized a small onboard computer to direct the spacecraft operations would be the most critical component they could design, and to test their ideas, they turned to the power of MIT's famous Whirlwind computer. This gigantic vacuum-tube computer was housed in an enormous building, and before turning Whirlwind on, the lab team needed to first notify the Cambridge power plant because of the tremendous strain the computer put on the city's electrical system.

With Whirlwind's help, the team figured out how to make it all work. The overall autonomous operation was managed on board by a small general-purpose digital computer, configured by its designer, Lab member Ramon Alonso. It didn't need much power except occasionally for higher-speed computations. A unique feature of this computer was a prewired, read-only, non-erasable memory called a core rope, a configuration using wires threaded in and out of tiny magnetic rings. A ring, or core, with wire threaded through the center represented a one; an empty core represented a zero. The pattern of wires formed the ones and zeros of a hardwired computer program.

The *Mars Probe* team's design was remarkable, their documentation comprehensive. In July 1959, they compiled a four-volume set of descriptions, details and schematics about the little spacecraft, the small computer and the guidance, navigation and control system. What the team didn't know at the time, however, was despite their groundbreaking work, their beloved *Mars Probe* would—sadly—never fly. But everything they designed, tested and calculated for this far-fetched little computer would soon transform into the guidance computer for the Apollo spacecraft.

In the fall of 1960, word reached Doc Draper that his good friend and former student Robert Seamans was going to join NASA.

"Just before I went down to Washington," Seamans recalled, "I got a call from Doc, and he said, 'Before you go down there and get involved in everything, how about spending a half a day over at the Lab so we can tell you what we're doing, that we think ought to be at least considered by NASA as part of the space program.'"

Seamans visited and was duly impressed. The concept for a self-contained computer and navigation system was intriguing. Seamans set up a meeting between Draper and Harry Goett to discuss how the Lab's ideas might fit into the various long-range plans Goett's committee had proposed for NASA, and in subsequent meetings, they determined the system should consist of a general-purpose digital computer with controls and displays for the astronauts, a space sextant, an inertial guidance unit with gyros and accelerometers and all the supporting electronics.

Charles Stark "Doc" Draper. Credit: Draper.

Hal Laning, Milt Trageser and Dick Battin with a model of the Mars Probe. Credit: Draper.

In all these discussions, everyone agreed the astronaut should play a role in operating the spacecraft—he should not just be along for the ride. And all the NASA people especially liked the self-contained navigation capability, since there was fear the Soviet Union could interfere with communications between a US spacecraft and the ground, endangering the mission and the lives of the astronauts.

From there, accounts diverge of what happened next, depending on who tells the story. Seamans says that after Kennedy's April 1961 challenge to land on the Moon, Seamans himself recommended the MIT Instrumentation Lab to NASA administrator James Webb but was worried about appearing to play favorites.

"I said, 'I know I'm prejudiced. I used to work in that laboratory. But if we want to have the most imaginative, innovative work done on that very key piece, I think it should be the Draper Lab,'" Seamans recalled. "And Jim Webb said, 'One of the important things when you're contracting is to know what you're going to get, and the best way of knowing is to know who the people are. You shouldn't look at yourself as being prejudiced. That's the kind of information we need.'"

And that's when, Seamans said, a sole-source contract for the Apollo Guidance Computer was offered to Draper and the Lab.

But Battin thought it was Robert Chilton, head of NASA's Flight Dynamics Branch at Langley, who recommended giving a contract to the Lab.

"He wrote a letter, which he gave me a copy of, recommending that we be included in the Apollo project," Battin said. "He said that the kind of work MIT was doing was really exactly what was needed for the Apollo program, because Apollo was not going to depend on communication with the ground, as the Russians might interfere with the communication link. I always thought that was the reason for our selection for Apollo."

David Hoag remembered it differently. In a paper he wrote years later, he outlined that several meetings took place in early to mid-1961 between NASA officials and Lab engineers to work out the details of a guidance computer for Apollo. By July, the lab had put together an eleven-page proposal that included just one graphic: a hand-drawn depiction of the Earth-Moon system. And on August 10, by letter, NASA contracted the Laboratory for the first year's development of the Apollo guidance and navigation system.

However, Doc Draper always insisted that Webb called him up on the phone to offer the contract.

Despite the differing stories, everyone agrees an approximate version of this terse discussion took place between Webb and Draper:

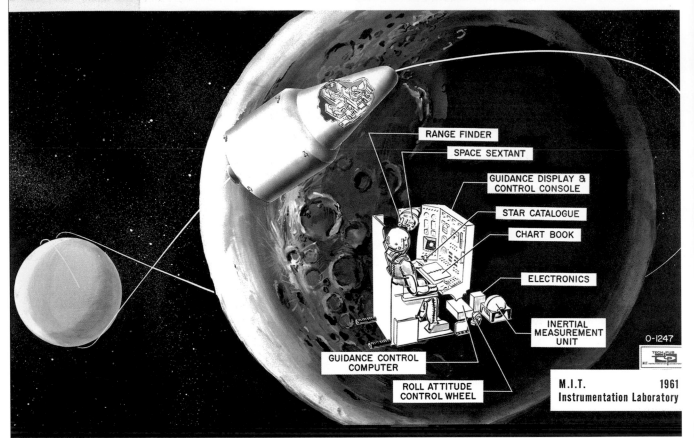

RANGE FINDER

SPACE SEXTANT

GUIDANCE DISPLAY & CONTROL CONSOLE

STAR CATALOGUE

CHART BOOK

ELECTRONICS

INERTIAL MEASUREMENT UNIT

GUIDANCE CONTROL COMPUTER

ROLL ATTITUDE CONTROL WHEEL

O-1247

M.I.T. 1961
Instrumentation Laboratory

This diagram depicts early ideas for the self-contained inertial-celestial navigation system for the Apollo spacecraft. The diagram details the layout of many of the system's component parts inside the Command Module, including the space sextant, guidance display and control console, star catalogue, chart book, inertial measurement unit and Apollo Guidance Computer. Credit: Draper.

Dick Battin at the Instrumentation Lab. Credit: Draper.

Webb: "Dr. Draper, you know we have to build a spacecraft to go to the Moon. And I think the guidance system is one of our hardest problems. Do you think you could help us with that?"

Draper: "Yes, of course."

Webb: "Well, then, when would the system be ready?"

Draper: "It would be ready when you need it."

Webb: "So, how will I know if it will work?"

Draper: "I'll volunteer to go along and fly it for you to the Moon."

However the details actually played out, the truth of the matter was that just weeks after Kennedy announced that NASA was to land a spacecraft on the Moon, the very first NASA prime contract for Apollo was signed with the MIT Instrumentation Laboratory to build the guidance and navigation system.

"We had a contract," Battin said, "but we had no idea how we were going to do this job, other than to try to model it after our *Mars Probe*."

Part of the lore of the Apollo Guidance Computer is that some of the specifications listed in the Lab's eleven-page proposal were basically plucked out of thin air by Doc Draper. For lack of better numbers—and knowing it would need to fit inside a spacecraft—he said it would weigh 100 pounds (45 kg), be 1 cubic foot (0.3 cubic meter) in size and use fewer than 100 watts of power.

Battin said a new guidance computer design they'd been tinkering with had 4,000 words of 16-bit memory, which was read-only, and 250 words of volatile memory, or RAM. This was small by any standard.

But at that time, very few specs were known about any of the other Apollo components or spacecraft, as no other contracts had been let. And NASA hadn't yet decided on its method—Direct Ascent, Earth-Orbit Rendezvous (EOR) or Lunar Orbit Rendezvous (LOR)—and the types of spacecraft to get to the Moon.

"We said, 'We don't know what the job is, but this is the computer we have, and we'll work on it, we'll try to expand it, we'll do all that we can,'" Battin said. "But it was the only computer that anybody had in the country that could possibly do this job . . . whatever this job might be."

While no other company had advanced in the computing area as much as the MIT Lab, there was a budding industry to create space-worthy navigation systems. So, the fact that no one else had the chance to bid on this potentially lucrative project created a controversy in the industry. But NASA stood by its choice. They also knew some of the other companies would have a chance for a piece of the guidance, navigation and control (GNC) pie.

From the outset, there was a clear understanding that MIT would do only the technical design and prototype development for the hardware. Following the precedent set by the Lab's work on the gunsights and missile-guidance systems, industrial contractors would conduct the manufacturing phase. Over the course of 1962 and 1963, NASA chose companies to produce specific parts: Raytheon would build the guidance computers, AC Spark Plug—part of General Motors—would manufacture the inertial guidance system and Kollsman Instrument Company would construct the optics for the system. The Lab would be responsible for creating the software.

The early conceptual work on the GNC proceeded rapidly, with Trageser, Battin and Laning working out the overall configuration. *Guidance* meant directing the movement of a craft, while *navigation* referred to determining present position as accurately as possible and how it related to the future destination. *Control* referred to directing the vehicle's movements, and in space the directions related to its attitude (yaw, pitch and roll) or velocity (speed and direction). MIT's expertise centered on guidance and navigation, while NASA engineers—especially those who had experience working on Project Mercury—emphasized guidance and control. The two worked together to create the many maneuvers that would be required based on the gyros' and accelerometers' data, as well as to determine how to make the maneuvers part of the computer and software.

NASA and the Lab at MIT decided the astronauts traveling to the Moon and back would need a telescope to periodically align the inertial guidance system to the stars, as well as to make navigation measurements with a sextant by observing the direction of the Earth and Moon against the background stars. The digital computer was required to handle all the data, and part of the computer would have a fixed, non-erasable, indestructible core-rope memory—an upgraded version from the *Mars Probe*. Another part of the computer would consist of erasable memory that would allow astronauts to input information. That meant the astronauts would need an arrangement of displays and controls to operate the system.

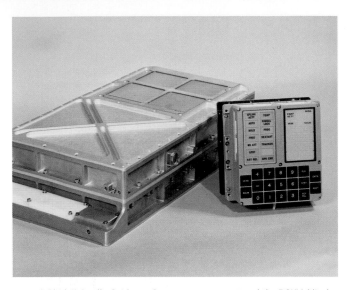

A Block II Apollo Guidance Computer component and the DSKY (display and keyboard). Credit: Draper.

Wernher von Braun visits the MIT Instrumentation Lab in March, 1964. Left to right: unknown, Milton Trageser, von Braun, Dick Battin and Ed Copps. Credit: Draper.

"Ramon Alonso was responsible for the core-rope memory," said Battin, "and he stopped me in the hall one day and he said, 'Hey, I've got a great idea of how to communicate with this computer.' He came up with a noun-verb communication, the design of the display panel and the keyboard, and he showed it to me. He said, 'What do you think of this?'"

Battin looked it over and said, "Well, it sounds good." He knew it would take some detailed, painstaking work to create a user interface that would integrate successfully into the computer.

With NASA's decision to attempt LOR, the Lab and the industrial contractor tasks were expanded to include the GNC for the Lunar Module. NASA chose Grumman Aircraft Engineering Corporation to build what was initially called the lunar excursion module; the name was later changed to just Lunar Module (LM) because the word *excursion* connoted a leisurely trip.

In early 1963, coordination meetings took place with Grumman for the LM and North American Aviation for the Command and Service Modules (CSM). The Lab decided on a self-imposed ground rule that the guidance computer hardware elements in each spacecraft should be as similar as possible. That decision would later pay off in manufacturing, testing and astronaut training. This ended up being one of the few systems that were the same on both the CSM and LM, although the software would need to be entirely different.

Knowing they would need to expand their operations to work on Apollo, the MIT Instrumentation Lab made arrangements to move into new facilities, a former underwear warehouse in Cambridge next to the Charles River. Battin thought they could maybe get by hiring 20 to 30 more people. What he couldn't possibly fathom in 1963 was that by 1968, the lab would need to employ approximately 350 people in order to complete their work for Apollo.

"It was a whole change in point of view, that this project was not really just a few of us working, like on the *Mars Probe*," Battin said. "It was a much bigger job, and we had to have lots of people whose only function seemed to be to keep everybody happy, not just NASA, but the spacecraft builders too. And we had to have firefighters to put out the fires."

Despite the Lab's experience with aerospace computers, the Apollo project quickly became a genuine challenge. It wasn't just the challenge of creating a complicated system, the likes of which had never existed before, but also creating the technology to make it possible. Fortunately, the early 1960s was a convergence of time when the digital age was just getting started and technological innovations were transforming traditional production. What the Lab couldn't find among the burgeoning digital industry they invented and developed. Battin and everyone at the Lab would need every bit (and byte) of advancement they could find.

Robert Seamans and Doc Draper. Credit: Draper.

In the meantime, Doc Draper made good on his statement that he would volunteer to fly to the Moon. He sent an official letter to Seamans, applying for the job as astronaut.

"I would like to formally volunteer for service as a crew member on the Apollo mission to the Moon, and also for whatever suborbital and orbital flights that may be made in preparation for the lunar trip," wrote Draper. "I feel that my work of the past thirty years has given me a unique background of well-rounded experience that may be considered as helping me to qualify for service as the engineering and scientific member of a spacecraft crew. It should be easier for me to learn the techniques of crew operations than for somebody else to acquire the capability I already possess. I believe I could do a good job for you."

But since Doc was well into his sixties, NASA had to decline.

"Doc Draper never let me forget it," Seamans recalled years later. "He said, 'You never acted on my suggestion. I was all ready to go to the Moon and you wouldn't let me go!'"

AT THE START OF 1963, JOHN PAINTER

knew that something had to change. He'd been in Houston at the Manned Spacecraft Center (MSC) since July 1962 and was certain that hardly anyone in his group knew what was going on. In five months, he had received no clear direction or encouragement from his supervisors, something he'd never experienced before. He was beyond frustrated.

But Painter was happy to be back in Texas. He'd met his wife, Joy, in 1954 when he joined the Air Force at Houston's Ellington Air Force Base. Following up with four years as an Air Force officer and navigator, and then four years of stretching his GI Bill dollars to squeeze in both a bachelor's and master's degree in electrical engineering from the University of Illinois, he was ready to start getting a paycheck.

Apollo astronauts during classroom training at the Manned Spacecraft Center, including Jerry Carr, Frank Borman, Jim Lovell, Buzz Aldrin, Neil Armstrong, and other unidentified persons. Credit: NASA.

Painter and Hondros arranged for lectures on particular kinds of radio hardware (such as antennas, transmitters and receivers), and Painter prepared a set of notes from which he would personally lecture on the theory of space communications. Visual demonstrations came in the form of hand-drawn paper flip charts on a three-legged easel. To make sure that no one would goof off, Painter and Hondros decided to make all the instructors write up their lecture in narrative form, so the astronauts could receive a set of study notes.

Painter realized his initial attempt at being an instructor wasn't stellar, but it was better than nothing. Some of the astronauts took it seriously (McDivitt, Borman, Armstrong). Some slept (White). And some cracked jokes (Conrad). John Young, still worried about bodily functions in space, commented about the proposal of live television from inside the spacecraft: "I don't care what anybody says. When I'm taking a crap, that damn TV camera is going to be off."

But during the course, Painter and The Greek got to know this group of astronaut pilots who would fly Gemini and then Apollo. The connections they made with those who would be operating the spacecraft—and the communications system—would soon pay off.

In the spring of 1963, things finally changed for Painter and Hondros. Barry Graves Jr. and Howard Kyle, who had both worked on setting up the Spaceflight Tracking and Data Network for Project Mercury communications, were now assigned to MSC. Graves had formed a new office called the Ground Systems Project Office (GSPO) to oversee the design and construction of what became known as Building 30, the Mission Control Center. Kyle was his assistant.

Graves was recruiting for the GSPO and had heard good things about the astronaut-training course in communications. He asked for a meeting with Painter and Hondros, inquiring if they would like to move over to his new organization. The two engineers looked at each other, smiled and told Graves they'd come if they could also bring the branch secretary, their friend Frances Smith, with whom they had developed a good rapport. The threesome became the first employees for the GSPO.

The functional requirements for the Mission Control Center were to be put together by Christopher Kraft, Glynn Lunney and their team of fellow flight controllers. The work of the GSPO was to compile requests for proposals, select the contractors from the submitted applications and monitor the contracts after they were awarded. Since neither Painter nor Hondros had the necessary civil service rank to be heads of any of GSPO's internal branches or sections, Graves hired a boss for the two engineers.

The GSPO crew moved over to better facilities at the Houston Petroleum Building, and while their new boss was taking two weeks off, Kyle gave Painter and Hondros an assignment. He wanted them to create a flight-test procedure for the Apollo communication system that had recently been selected (in December 1962) called the Apollo Unified S-Band Telecommunications and Tracking System. Painter and Hondros didn't know anything about it, and Kyle wanted the test procedure delivered in just two weeks.

The two engineers gathered what information they could on the system and, wanting to succeed in their test-procedures task, worked twelve hours a day for fourteen days straight. At the end of the two weeks, they were so exhausted, they were punch-drunk and jumped on top of their new boss's desk to dance a celebratory jig. But they delivered the procedure on time.

"It wasn't very good," Painter said, "but it seemed to satisfy Mr. Kyle."

After their initial assignment to write the flight-test procedure, Painter and Hondros began pursuing in earnest the details of the Apollo communications system. Apollo required additional bandwidth to uplink and downlink more data—but with Kennedy's deadline of getting to the Moon by the end of the 1960s, there just wasn't time to create an entirely new system. NASA turned to a design built by the Jet Propulsion Laboratory (JPL) for tracking and communicating with their robotic missions to deep space. JPL's system was technically called the Block III Receiver Exciter and Mark I ranging system, but the common nomenclature among those who worked in the industry was the Unified S-Band system (USB).

The USB system was an elegant and very capable solution for the communications challenges posed by human spaceflight to lunar distances, and it combined voice communications, telemetry, command, tracking, ranging and television into a single transmitter system. It also saved size and weight—significant for getting a weight-conscious spacecraft to and from the Moon—and simplified operations.

"Digital data was really something quite new," Painter said, "and the USB system would transmit many measurements on the spacecraft and health status of the astronauts back to Earth. And it would transmit digital data from the controllers up to the astronauts, for entry into their onboard flight computer. It would even handle the transmission of live TV from the spacecraft to the ground. Perhaps most importantly, the radio system would transmit a digital data stream, called a range code, up and back to track the path of the spacecraft, from which its precise trajectory could be calculated in the ground computer."

But the system hadn't been completely designed yet with the new modifications for Apollo, and the people in the Flight Operations Division hadn't yet developed a set of operational requirements for all the kinds of data they wanted to transmit to and from the Apollo spacecraft. When they looked at some of the paperwork submitted by the contractors for the Apollo CSM and the LM, Painter and Hondros noticed discrepancies in how the communications systems would be incorporated into the two spacecraft. They came up with a plan of action to understand the system well enough to write a set of technical reports so that everyone would have the same documentation, specs and requirements. This would keep everyone on the same wavelength, so to speak.

"Everyone" included the prime contractors for the CSM, North American Aviation in Downey, California; the contractors for the LM, Grumman Aircraft Engineering Corporation in Bethpage, New York; the team developing the Saturn booster rocket at Marshall Space Flight Center (MSFC) in Huntsville, Alabama; and several divisions at MSC, all of which would use the radio system. The ground equipment for the communications stations around the world was being coordinated by the Goddard Space Flight Center in Maryland.

In early 1963, none of these companies and agencies knew what the modified Apollo design would be. Only JPL could predict that. So, Painter and Hondros started traveling to the different entities involved, going first to California to talk with the JPL people who had designed their deep-space radios.

JPL is nestled in California's San Gabriel Mountains, and Painter and Hondros found the weather, landscape and nightlife to be delightful. They worked out of a $9-per-night motel on Pasadena Boulevard, driving a rental car up to JPL. At night, the two explored Los Angeles and Hollywood, looking for movie stars. Every place along Hollywood Boulevard where they ate, Hondros would discover a fellow Greek from the "old country," and some of the meals ended up being free. Once, Hondros ran into a former high school friend, who was playing in a nightclub band.

At JPL, Painter and Hondros tracked down two of the original analysts and designers of the USB system. One was an "old guy," age forty-five. Painter had been trying to unravel the mathematics of the system from a report done by one of the Apollo contractors. The old guy took a look at the contractor's mathematics, which ran several pages. He then went to the blackboard and derived the sought-after result in three lines of equations.

"John, when something looks that complicated, you're probably looking at it the wrong way," the old guy said, something Painter would always remember.

The two engineers returned to Houston to document in their own report of the mathematics of the USB as it was to be redesigned for Apollo. Having the mathematics, the performance of the USB system could be calculated, channel by channel, using MSC's computers.

Painter and The Greek then began the task of writing their "NASA Technical Notes." The first volume would physically describe the complete operation of the Apollo Unified S-Band Telecommunications and Tracking System, what it should do and how it should work. The second and much more ambitious document would be a complete mathematical model of the system, which could be used to compute numerical predictions of how well the system would work; it would give what are called numerical margins, above or below that needed to provide the desired performance of all the various system channels. A third volume would include a numerical tabulation of system performance, channel by channel, for all three spacecraft.

But Painter and Hondros's plan to get everyone on the same page ran into some snags.

"The technical problems were thorny enough, but there were political problems also," Painter said. One issue was that North American Aviation was not communicative—or forthcoming, Painter thought—about what they were doing.

"Another problem was," Painter said, "MSC had responsibility and authority only for the Apollo spacecraft. But this system [the USB] contained ground stations. And people in Houston did not have any authority over design of the ground stations. Legally, only Goddard Space Flight Center in Maryland had that responsibility."

But the system had to be designed as one entity. That is, both the spacecraft and ground elements of the system had to be developed together to make sure that they would be compatible with each other. And nobody in NASA had the responsibility to ensure that compatibility.

So The Greek and Painter took on the responsibility—with Graves's backing, since he would bear the ultimate responsibility—of ensuring compatible design of all the parts of the Apollo Unified S-Band System.

Even though everyone at MSC was still scattered among the various buildings along the Gulf Freeway, Painter and Hondros became aware of how the many branches and divisions of MSC would have to work together, especially since so many different systems and components for Apollo needed to work together.

An aerial view of the Manned Spacecraft Center, in 1963 during early construction. Credit: NASA.

THE MSC ITSELF WAS COMING TOGETHER,

with construction continuing on many unique facilities, some designed with the help of Henry Pohl. He had been recruited to help design the test facilities for MSC since he had been heavily involved in all the testing in Huntsville and knew there were several things he would have set up differently, given the chance. In his mind, an engineer who couldn't properly test their theories wasn't worth a damn; so, when he came to Houston, he was ready to assist with some of the MSC designs. The idea was to incorporate all the test facilities in Houston so they could do on-site assessments of every system and subsystem needed for human spaceflight. In the early 1960s, scientists and engineers had a limited understanding of the effects of space on hardware and humans, so, being able to replicate the environment of space as much as possible was vital.

Pohl had actually seen some of the initial plans for facilities at MSC when he visited Langley in 1961 to discuss his possible move to Houston. When he first saw the plans for how the entire MSC would be laid out, Pohl got the impression from some people at Langley that the main facilities for offices and other buildings were going to be set up like a university: If this whole going to the Moon thing didn't pan out, Rice University could have a second campus. He also overheard conversations that some people weren't very keen on the idea of moving to Houston because it might be career suicide. Pohl didn't quite know what to make of it.

Once in Houston, he met with Dick Ferguson to review some of the plans for the test facilities. Pohl looked intently at the schematics for a vacuum chamber, and he recalled a time he had tested a small rocket engine at the chamber in Huntsville. He had placed the rocket too close to the floor of the chamber, and when he fired it, all the exhaust hit the floor, bounced right back up and killed all the vacuum around the engine—which meant he didn't get the data he wanted. Plus, it burned the test stand and everything around the engine. After contemplating the issue, he fixed it by putting in a smooth deflector to channel the exhaust away from the engine.

Construction of the large Chamber A in the Space Environment Simulation Laboratory in 1964. Credit: NASA.

He saw the MSC chamber design was configured for a test stand that would allow an engine to fire right up against a wall.

"Dick, you can't do that," Pohl told Ferguson. "That exhaust is going to come back and hit that engine in five milliseconds, and you're not going to get any data."

Ferguson said, "Henry, don't make those accusations unless you can verify them."

"Well, that's easy to do," Pohl said. "It's the square root of KGRT, with K as such and such; G and R are these values; T is such and such; and it's the square root of that."

Ferguson was good with figures. He took out his pencil and quickly did the math on the side of the design paper. Contemplating the numbers, he said, "Hmm, you're right."

"So, we went and redesigned the vacuum chamber," Pohl said. "He just thought I was brilliant because I knew that. I never did tell him that I had been down that road, had tried and made that mistake and learned the hard way."

Pohl continued to help with designs with the Space Environment Simulation Laboratory (SESL), which would include two vacuum chambers—one big enough to put spacecraft inside. The interior of Chamber A would be a 90-foot (27-m)-high, 55-foot (17-m)-diameter stainless-steel vessel that could simulate pressures and temperatures equivalent to 130 miles (209 km) above the Earth. The interior walls could be cooled to −280°F (−173°C) to re-create the icy conditions of space, and two banks of carbon arc lights would simulate the unfiltered light and heat from the sun. In addition, the chamber would feature a rotating floor so the objects placed inside could experience orbital and flight variants. So, except for weightlessness, NASA would be able to reproduce most of the conditions the spacecraft would encounter in Earth orbit or on flights to the Moon, with the added advantage of being able to return spacecraft and crew to Earth's atmospheric conditions in a matter of seconds.

Chamber B was designed to be smaller—26 feet (8 m) in height and 25 feet (7.5 m) in diameter—and would be used for testing smaller systems like the spacecraft environmental control systems and spacesuits, as well as allowing astronauts to train in vacuum conditions.

While vacuum chambers—sometimes called altitude chambers—had been used for decades, previously, nothing had been designed on such a large scale to accommodate the larger Apollo spacecraft. And hardly any chambers had been built to be human-rated, meaning they had to be designed with airlocks and holding areas for rescue personnel, emergency repressurization systems, and biomonitoring and surveillance systems as safeguards for the people who would enter. The SESL would require cutting-edge technology.

James McLane Jr. was instrumental in helping design many of the test facilities for MSC. In the 1940s, he worked at Langley for the National Advisory Committee for Aeronautics (NACA), and in 1951 he moved to Tullahoma, Tennessee, to design wind tunnels and other facilities for the Army Corps of Engineers and the Air Force. He joined NASA in Houston in 1962.

Right: Dr. Wernher von Braun stands by the five F-1 engines of the Saturn V launch vehicle, mounted on the Saturn V S-IC (first) stage. The engines measured 19 feet (5.7 m) tall by 12.5 feet (3.8 m) at the nozzle exit and burned 15 tons of liquid oxygen and kerosene each second to produce 7,500,000 pounds of thrust. Credit: NASA/Marshall Space Flight Center.

Exterior view of the Vibration and Acoustic Facility, Building 49, in 1965. Credit: NASA.

"Ken saved the day," said Wren. "He's one of the smartest people I've ever had the pleasure of knowing and working with. He could do this very deep, detailed technical work, but he also had the ability to convey that to idiots like me, the laypeople, so we could understand all the equations!"

Creating the right sound for the acoustics wasn't the only problem; they had to design a building that could withstand the sound waves and vibrations too.

"You cannot put the test article in a huge space chamber or a 'reverberation room,'" Wren said. "First, the walls of such a room would have to be very strong and resistive to a dynamic forcing function, as it might couple with the natural resonant response frequency of the wall and vibrate it to failure. Also, standing waves would be created inside a hard-walled chamber, and this would not apply a dynamic forcing function to the test vehicle in the appropriate manner to simulate launch and boost."

For aesthetics, the architects of MSC wanted the exterior of any facility to be uniform in design, so the building needed to include the same exterior panels as on all the other buildings (which would also lower the costs by mass-producing the exterior panels). The panels were called pre-cast exposed aggregate facing concrete panels. So, the design for this new test facility would have to make the concrete panels work.

Eldred came up with a clever approach to solve this dilemma, helping Wren and his team design and create what is called Building 49 at MSC: the Vibration and Acoustic Test Facility, also known as the Twin Towers.

"His approach was to cloak the test article in a series of massive shrouds that completely enveloped the vehicle," said Wren. "Massive high-intensity acoustic drivers would be positioned on top of each of these shrouds, and carefully controlled acoustic sound waves would be sent down over the vehicle from these drivers to excite the natural structural responses of the vehicle."

The acoustic forcing functions were carefully chosen relative to frequency and amplitude to duplicate the Saturn V–generated noise at lift-off from the launchpad and the noise impingement experienced when the vehicle went through maximum dynamic pressure.

"We put it in the back of the MSC site," Wren said, "and the design looks kind of funny when you see it because it has some low buildings and then a couple of towers." The facility they designed had about 13,700 square feet (1,273 sq. m) in total, with space for offices and control rooms, and the two test towers were 42 feet (13 m) and 115 feet (35 m) tall.

One tower was for vibration—researchers could put in a spacecraft and shake it in "free" mode or they could tie it down and do a fixed-base mode, depending on what kind of conditions they needed to test. The shaker system could shake up to 10,000 pounds (4,536 kg).

"The other tower would be where we were going to put great big old acoustic horns in there, like your speakers for your hi-fi or your stereo equipment," Wren said. "But these were huge, huge power amplifier drivers, so we had some problems to solve with that."

Like everything with Apollo, the design had to be done quickly.

"The environment was hectic, to say the least," Wren said. "We had a big job to do and we weren't quite sure how to do it. But we were lean and mean and going around the clock 24/7, and we weren't encumbered with a lot of paperwork. A lot of us came from the military, so we knew how to work with preciseness and keeping up with parts and testing the parts and equipment, but we didn't have to go through a lot of paperwork rigmarole. Our administrator, James Webb, he worked with Congress and took care of all the money aspects and that allowed us in the technical world to just go like crazy and get the job done."

A Saturn rocket on the Mobile Launch Platform outside the newly built Vehicle Assembly Building at Kennedy Space Center in 1966. Credit: NASA.

CONSTRUCTION WAS UNDER WAY AT A frenetic pace at the other NASA centers involved with Apollo as well. The launch complex at Cape Canaveral in Florida grew from its origins as the Missile Firing Laboratory under the Army Ballistic Missile Agency (ABMA) in Huntsville, and when the ABMA was usurped by NASA to become the Marshall Space Flight Center (MSFC), the launch facilities in Florida became the Launch Operations Directorate. In 1962, with expanding operations in the buildup to Apollo, it received the status of a full-fledged center, called the Launch Operations Complex (LOC).

Two launchpads were constructed for the Saturn I and IB rockets at the northernmost end of the site, LC-34 and LC-37. The new facilities were on Merritt Island, immediately north of Cape Canaveral and the Air Force station and included a launch control center, and a 130-million-cubic-foot (4-million-cubic-meter) Vertical Assembly Building (VAB). This monstrosity, still under construction in 1963, measured 525 feet (160 m) tall by 716 feet (218 m) long by 518 feet (158 m) wide, with the largest interior volume in the world. More launchpads were required to handle the giant Saturn V for sending spacecraft to the Moon, so Launch Complex 39, with pads A and B, was being erected. In addition, NASA had plans to build at least two more launchpads.

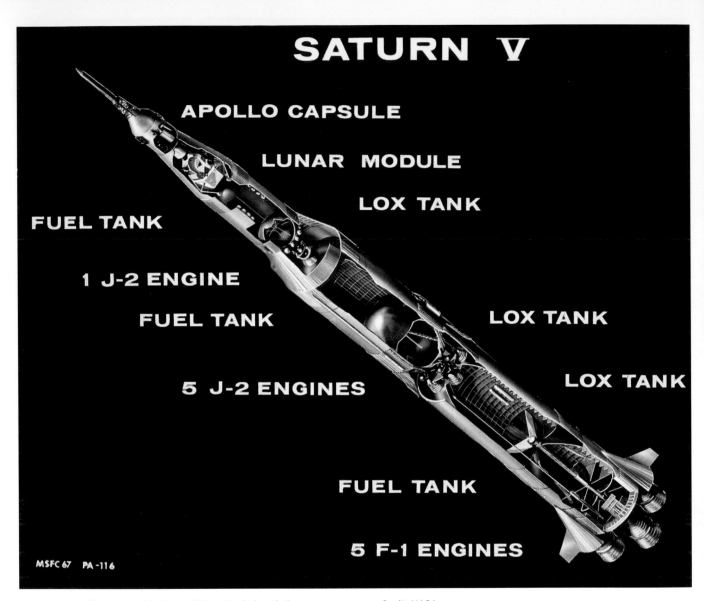

SATURN V

APOLLO CAPSULE

LUNAR MODULE

LOX TANK

FUEL TANK

LOX TANK

1 J-2 ENGINE

FUEL TANK

LOX TANK

5 J-2 ENGINES

FUEL TANK

5 F-1 ENGINES

MSFC 67   PA-116

A cutaway illustration of the Saturn V launch vehicle with the major components. Credit: NASA.

The rockets would be assembled and joined with the spacecraft on a mobile launcher platform inside the VAB and then moved on a giant mobile transporter to one of the launchpads. Also under construction were two vacuum chambers for tests and an Operations and Checkout Building where the Gemini and Apollo spacecraft could be received before being joined to their launch vehicles. Kurt Debus, a member of von Braun's original V-2 rocket engineering team, was named the LOC's first director. He led the design, development and construction of the center.

Von Braun remained the head of his rocket-building team and director of MSFC, where a rapid buildup of facilities was also taking place. The main headquarters, the nine-story Central Laboratory and Office Building (irreverently called the von Braun Hilton) opened in July 1963. Since MSFC was NASA's lead center for the development of rocket propulsion systems and technologies, other facilities being built included the Dynamic Test Tower to perform shaking and vibration tests, and a giant, 405-foot (123-m) high static test stand capable of handling boosters up to 178 feet (54 m) in length and withstanding a thrust of up to 7.5 million pounds (3.4 million kg) in order to test-fire the Saturn V booster.

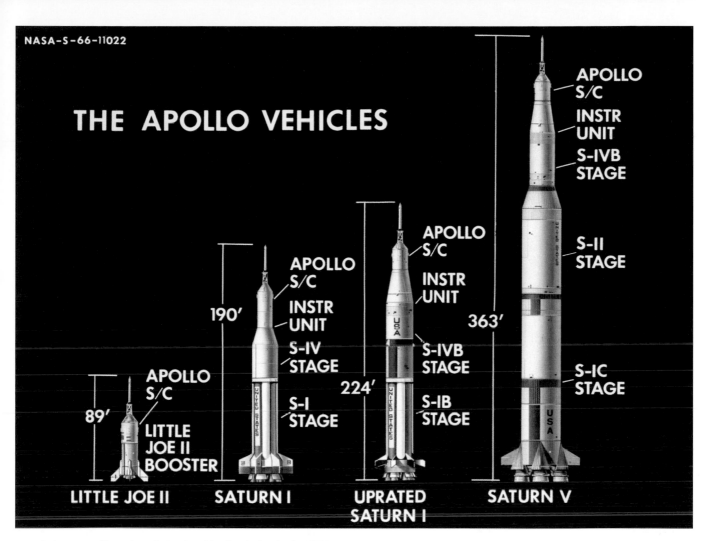

THE APOLLO VEHICLES

89' — LITTLE JOE II
APOLLO S/C
LITTLE JOE II BOOSTER

190' — SATURN I
APOLLO S/C
INSTR UNIT
S-IV STAGE
S-I STAGE

224' — UPRATED SATURN I
APOLLO S/C
INSTR UNIT
S-IVB STAGE
S-IB STAGE

363' — SATURN V
APOLLO S/C
INSTR UNIT
S-IVB STAGE
S-II STAGE
S-IC STAGE

Artist concept illustrating relative size of Apollo vehicles. Credit: NASA.

The engineering required to leave our planet resulted in a construction and manufacturing boom across the country, along with the development of new technologies that previously had never even been considered. Although the Saturn V was developed in Huntsville, all the various stages and pieces were constructed by multiple contractors.

"Understandably, the entire aerospace industry was attracted by both the financial value and the technological challenge of Saturn V," von Braun wrote about the project. "To give the entire plum to a single contractor would have left all others unhappy. More important, Saturn V needed the very best engineering and management talent the industry could muster. By breaking up the parcel into several pieces, more top people could be brought to bear on the program."

The Boeing Company was the successful bidder on the first stage of the Saturn V (S-IC), North American Aviation won the second stage (S-II) and Douglas Aircraft fell heir to Saturn V's third stage (S-IVB). The inertial guidance system had emerged from a Marshall development in-house, creating what could be considered the big rocket's central nervous system, called the Instrument Unit (IU), housing all the electronic gear to make the rocket function. IBM produced the launch vehicle computer, and RCA was tasked with building nineteen ground-computer systems to be used in the checkout, static test and launching of Saturn IB and Saturn V vehicles at Marshall.

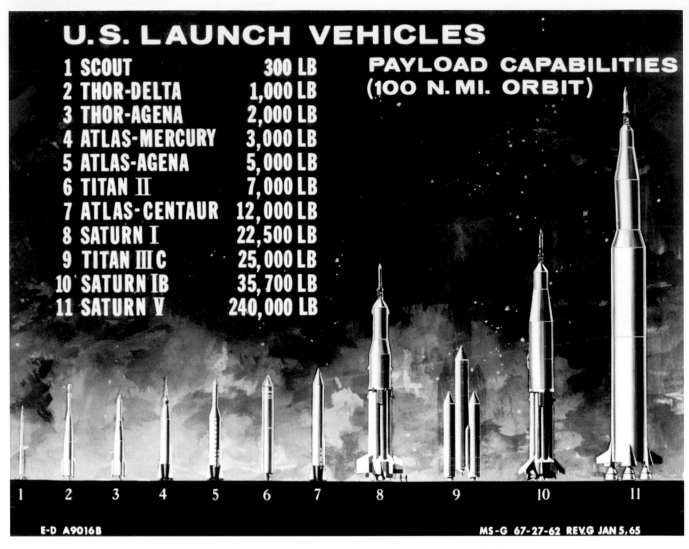

U.S. LAUNCH VEHICLES

PAYLOAD CAPABILITIES (100 N. MI. ORBIT)

#	Vehicle	Payload
1	SCOUT	300 LB
2	THOR-DELTA	1,000 LB
3	THOR-AGENA	2,000 LB
4	ATLAS-MERCURY	3,000 LB
5	ATLAS-AGENA	5,000 LB
6	TITAN II	7,000 LB
7	ATLAS-CENTAUR	12,000 LB
8	SATURN I	22,500 LB
9	TITAN IIIC	25,000 LB
10	SATURN IB	35,700 LB
11	SATURN V	240,000 LB

E-D A9016B

MS-G 67-27-62 REV.G JAN 5, 65

Artist concept of the various launch vehicles used by the United States and their payload capabilities. Credit: NASA.

But the giant rocket soon grew too large and complex for just one facility. New sites were needed, and engineers realized that some of the sites needed to be far away from populated areas yet near easy transportation routes. Waterways provided the simplest means of transportation, so several sites were chosen: the Michoud Assembly Facility at New Orleans, an old sugar plantation used for military manufacturing during World War II and the Korean War; the Mississippi Test Facility at Bay St. Louis, Mississippi; and the Slidell Computer Facility at Slidell, Louisiana. Then there were the NASA Rocket Engine Test Site at Edwards Air Force Base in California and the production facilities stage at Seal Beach, California. Apollo was truly becoming a nationwide enterprise, and the increase in facilities meant new roads, railway lines and shipyards were needed (not to mention special barges, ships and airplanes).

The Saturn rocket itself was a complicated, massive beast needing a variety of new technology and cobbled-together previous parts and pieces, which came in a jumble of confusing names and acronyms. Originally, in the late 1950s and early 1960s, the concept for Saturn was to have a family of multimission rockets in various configurations and names, since at first it wasn't clear what the Saturns were supposed to do. When Kennedy chose the Moon, and when LOR was chosen in 1962, the path became clearer—the family of rockets was reduced from nearly a dozen configurations (with names such as the Saturn C-1, C-2, C-3 and so forth) to three. To avoid any confusion regarding nomenclature, in February 1963, the NASA Project Designation Committee suggested renaming the three to Saturn I, IB and V.

The two-stage Saturn I and IB rockets were developed to test some of the Saturn V hardware, perform preliminary flight tests and send up the first Apollo astronauts into Earth orbit. The difference between the two configurations was that the IB was slightly taller, to carry more propellant so it could lift a more massive load. These rockets were based on the earlier Juno and Redstone rockets.

Saturn V would consist of three stages: The first, called S-IC, was powered by five F-1 engines—the largest rocket motors ever built—burning kerosene and oxygen. The second stage, S-II, would need about a million pounds of thrust, requiring the development of new 200,000-pound hydrogen-oxygen engines called J-2. The third stage, S-IVB, would rely on a single J-2 engine to boost the Apollo CSM and the LM, as well as the "brains" of the rocket, the IU. That same S-IVB stage would serve as the second stage for the Saturn I and IB rockets.

Apollo needed one final kind of rocket, since the Saturn would require a launch-escape system for the astronauts in case of a launch failure. General Dynamics, Convair division, was selected to provide the relatively small and aptly named Little Joe II rockets to pull the Command Module (CM) away from the rest of the Saturn stack in the event of a life-threatening emergency.

LAUNCHES FOR THE MERCURY PROGRAM

came to an end with Gordon Cooper's Mercury 9 mission, which launched on May 15, 1963. This was the only "long-duration" flight of Mercury, lasting thirty-four hours, nineteen minutes, forty-nine seconds, with Cooper completing twenty-two orbits to evaluate the effects of a full day and more in space. Cooper's flight plan had him conducting Earth observations and photography—he was tasked with looking for potential Soviet nuclear launch sites among other things—collecting urine samples and monitoring his ship's status.

All went well until the twenty-first orbit, when he lost most of the electrical power inside the spacecraft. Cooper also noticed the carbon dioxide level was rising in the cabin and in his suit. As he realized multiple systems were failing, the low-key Cooper only commented to one ground controller, "Things are beginning to stack up a little." When it was time for deorbit, astronaut John Glenn talked Cooper through the steps, counting down to retrofire over the radio. Cooper hit the button—which didn't light up—so the only confirmation it worked was the change in velocity he could feel as the three small engines ignited behind him. Ground controllers could see the retro-rockets were fired on the mark. Glenn reassured Cooper, "Right on the old gazoo. It's been a real fine flight. Real beautiful. Have a cool reentry."

Even if the reentry wasn't cool, it was nearly perfect. The automatic reentry system wasn't working, so Cooper had to control the spacecraft's motions by hand during descent and manually deploy both his drogue and main parachutes. *Faith 7* landed within sight of the recovery ship, the USS *Kearsarge*, in the Pacific Ocean.

With that, the Mercury program ended and was deemed a remarkable success. Everyone—flight controllers, communications teams, spacecraft engineers, launch technicians, astronauts and more—gained experience on each flight. As the work moved toward Gemini, the concepts were further developed, expanded and improved—all essential steps to get to the Moon.

An additional Mercury flight, a proposed three-day extended Mercury mission, wasn't to be. The spacecraft, *Freedom 7-II*, would have been flown by space veteran Alan Shepard, but there had been rumors for quite some time this flight might be cut. Several NASA officials, including administrator James Webb, argued that Gemini was already primed for long-duration missions, and it was pointless to demonstrate a capability just once with a system about to become obsolete. Moreover, an accident on Shepard's flight could set the whole space program back, which NASA couldn't afford. In mid-June, the flight was officially canceled and the spacecraft put into storage. Everyone in the agency could now focus their efforts on the upcoming Gemini and Apollo missions.

AT THE SMALL AIRPORT IN CEDAR RAPIDS,

Iowa, two men strode out of a DC-4 airplane, down the mobile stairway and onto the tarmac. All eyes turned in their direction. Their black, well-tailored, fine-line gabardine suits, Stowe-Ivy collared shirts and colorful ties weren't often seen in these parts. They were outfitted with dark sunglasses and rich leather briefcases. People around them wondered if they were possibly FBI, or maybe even CIA. But then they noticed the shoes. One of the men sported flamboyant, bright yellow patent-leather dress shoes. Definitely not CIA. Could they perhaps be movie stars? Here in Iowa?

But now, what was this? One of the men—the one without the yellow shoes—was approached by a small group of people, and there were handshakes all around, even a few hugs. Family? If he were a movie star from Cedar Rapids, he certainly would be known. But none of the gawking onlookers recognized him or his family. Curiosity abounded. Who were these guys?

In the summer of 1963, these two well-dressed men were part of the fast-paced, high-demand, high-stress, high-paying world of contract engineers. In the dawn of the space age, electrical engineers were a sought-after commodity. Contract engineers moved from company to company, from military base to government installation, wherever the demand and dollars were, especially for engineers with the rare skills these two possessed.

"I grew up in vacuum-tube technology but also am a transistor expert," said Earle Kyle, the engineer without the yellow shoes. "So, I'm not only an analog and a digital guy but also a vacuum-tube and solid-state microchip guy."

Because the US was so desperate for electrical engineers, the joke was, Kyle said, if you could fog a mirror and hold a soldering iron, you were a wanted man, even if you didn't have a college degree.

"Even two years before I graduated, the phone was ringing off the hook," he said. "Somehow they tracked me down, and I got lots of job offers, with salaries that were crazy."

Kyle was tempted by the offers but had just gotten married, and his wife was an intellectual who insisted he not follow temptation and get his degree. In the end, it was the right thing to do, but Kyle often thinks about the money he missed out on during those years.

Originally from Minneapolis, Kyle had been living in Los Angeles, working several contract-engineering jobs for various military installations. Most recently, he had been in Goodyear, Arizona, working on the classified SR-71 Blackbird spy plane. That's where he met his yellow-shoed compatriot, Paul.

"We worked hundred-hour weeks—it was a crash program, making big bucks," Kyle said. "We didn't get any benefits, but they paid us cash in envelopes at the end of the week. Since Paul's last name is similar to mine, they always got us mixed up, which ticked me off because I was making more than him."

Kyle is hesitant to provide Paul's last name, as there still may be warrants out for his arrest; at the very least, he probably has a number of unpaid speeding tickets. During the Blackbird contract period, Paul rented a luxury home in Scottsdale from a professional baseball coach who never lived there in the summer. The house had an indoor-outdoor sunken pool and a beautiful view of Camelback Mountain out the front window. Paul lived the life of a playboy bachelor, with lovely looking girls seemingly always on one or both arms. He hosted parties every Wednesday night, which Kyle said were legendary across the nation in the contract-engineering world. There may have been times the police were involved.

Kyle commuted from LA to Phoenix, and during the week he stayed with Paul.

"Paul drove a Jaguar XKE convertible—always with the top down—and it seemed like we were constantly late to work," Kyle smiled. "This was all government secret stuff, with chain-link fences. They closed the main gate at 8 a.m. sharp, and if you were a minute late, you had to go to a different parking lot and walk way around to get in. So, we'd be racing down the highway from Phoenix into the desert at 120 miles [193 km] per hour, with cops chasing us. But they could never catch us with that V-12 engine."

Their latest contract had Kyle and Paul on a six-month stint at Collins Radio Company in Cedar Rapids, working on specialized communications equipment for Apollo. While Paul was dismal about the party scene in Iowa, Kyle was ecstatic. He was really, truly working on something for NASA and the space program. Since he was young, he had been captivated by the idea of exploring space. As a science fiction nut for as long as he could remember, he'll never forget the day during his eighth-grade year when his parents' *Collier's* magazine arrived.

Left: Astronaut Gordon Cooper leaves the Faith 7 *spacecraft after a successful recovery operation. The MA-9 mission, the last flight of the Mercury Project, was launched on May 15, 1963, orbited the Earth 22 times and lasted for 1½ days. Credit: NASA.*

Martin Luther King, Jr. speaking at the Civil Rights March on Washington, DC, on August 28, 1963. Rowland Scherman, Photographer. National Archives.

"The cover story had Chesley Bonestell paintings of spacecraft going to explore Mars and the solar system," Kyle recalled. "It just knocked my socks off, it just grabbed me."

For Kyle, space exploration is what he calls "the tuning fork of the soul"—where, if you're lucky, something grabs you so hard that you know it's what you want to do for the rest of your life.

"I was always talking about space," Kyle said. "My parents were supportive but my teachers and friends laughed and said, 'Forget about that crap, nobody is going to send things into space, not to mention people, so go do something practical with your life.' But after *Sputnik*, it wasn't so far-fetched anymore. Suddenly I wasn't a kook anymore."

So, even though working in Iowa wasn't glamorous, it was Kyle's open door to the real world of space travel. Collins Radio had built the communications systems used for the Mercury spacecraft and now they were working on the Apollo USB radio systems for the CSM and LM. They also were creating the voice-communications headsets that would be placed in the spacesuit helmets so the astronauts could talk directly with Mission Control.

Collins Radio paid for Kyle and Paul to fly to Iowa, as well as for their temporary housing. "I think they were kind of spooked about guys like us," Kyle said, "so they wouldn't let us in the main factory area. Instead, they put us in rented offices in the basement of the Younkers department store in a shopping mall down the street. I think they wanted to keep us away from the country bumpkins so we didn't contaminate them."

Their contract had them working on the helmet headsets, and the two engineers quickly figured out the issues that Collins had hired them to fix. "They thought the work was more complicated than what it really was, so as time went on, we started coasting, just letting the rest of the contract run out," Kyle said. "And as a consequence, we started getting sort of mischievous."

It was a hot and humid summer that year in Cedar Rapids. Down in the bowels of the shopping center, it was steamy, and dumpsters with rotting garbage from nearby restaurants attracted cockroaches.

"These cockroaches were huge," Kyle laughed, "and we'd catch them, stick a wad of chewing gum on them and stick in a toothpick with a handwritten banner attached with various obscenities. There was a bank of typists just down the hall, working on typing up manuals for various equipment, and we'd let the cockroaches go and take bets on how long it would take to hear screaming."

As thrilling as it was to be working on equipment for the space program, Kyle found the contract-engineering life tiresome. He missed his wife and their baby out in LA, he was worried about his mom and he was concerned about the racial situation in the US.

As he watched the news every night on the fuzzy black-and-white television screen in his Cedar Rapids apartment, it seemed as though each newscast would bring a new dramatic turn to the racial tensions permeating the country and the world in 1963. There was Alabama governor George Wallace standing on the steps of his state capitol declaring, "Segregation now, segregation tomorrow, segregation forever," and later, he literally stood his ground to block three black students from registering at the University of Alabama. Then, during a protest in Birmingham, policemen turned dogs and high-pressure water hoses on children as young as six. As the summer wore on, there were hundreds of demonstrations across the country, both for and against civil rights. From marches to mass meetings to nonviolent sit-ins at Woolworth's lunch counters, people were enduring physical attacks and getting arrested for their beliefs. Some were getting killed.

On the evening of June 11, President Kennedy addressed the nation on the issue of civil rights, for the first time roundly condemning segregation and announcing his intention to submit a comprehensive civil rights bill to Congress. And the very next evening, Medgar Evers of the NAACP was assassinated in his driveway in Jackson, Mississippi.

Marches and protests continued throughout the summer. A march for jobs and freedom was scheduled for late August in Washington, DC, and Kyle's mother was planning to attend with a delegation from the Twin Cities.

"She was a civil rights proponent, and she was getting heavily involved in the movement," Kyle said. "I talked to her on the phone, trying to stop her from going as I was afraid there would be violence, as had happened in Birmingham. But she was determined to go."

On August 28, Kyle took part of the day off from work at Collins to watch the television coverage of the march. He saw clips on the news of how the event drew about 250,000 people. And Martin Luther King Jr. gave a speech. In one piece of news footage, Kyle thought he could just make out his mother's face on the left front side of the reflecting pool in front of the Lincoln Memorial.

As he watched the TV, Kyle reflected on his heritage. "My great-grandfather had been a slave in Claiborne, Alabama; and my dad was born before the Wright brothers took man's first flight. So here I was, one of the few black aerospace engineers in the country, on the threshold of helping humanity go to another celestial body, and yet we still had racial hate."

AS 1963 PROGRESSED, THE ENVIRONMENT

in the young NASA organization was unsettled. From its inception, the agency had been asked to grow quickly and do extraordinary things. But as work proceeded on the rockets and spacecraft, a number of technical problems unveiled themselves, promptly leading to scheduling issues—the timelines for Gemini and Apollo started to slip. It was becoming apparent NASA might not get to the Moon before 1970 unless things changed. Clashes emerged between management styles in the organization's upper echelons. In the meantime, Congress decided to cut NASA's budget.

The primary technical issue came with the F-1 engine, which would power the first stage of the Saturn V rocket needed to launch the Apollo spacecraft into Earth orbit. During a test firing of an F-1 in late 1962 at Edwards Air Force Base in California, the engine destroyed itself. The cause, after detailed assessment, came from what rocket engineers call combustion instability, meaning the propellants—in this case, RP-1 (a type of kerosene) and liquid oxygen—weren't properly flowing together to produce the correct burn, which in turn meant the engine had to be redesigned. Von Braun's team worked with the contractor, the Rocketdyne Division of Rockwell International, to rejigger the 2,500-pound (1,134-kg) turbopump that was supposed to pump in the propellants at 42,500 gallons (161,000 L) per minute. Uneasiness prevailed in the intervening months, and the redesigned F-1 wouldn't be delivered to MSFC for more testing until October 1963.

While NASA's budget had been on the increase in 1961 and 1962, Congress decided in 1963 to cut the agency's budget by 10 percent. The widespread political and public support for the lunar initiative—buoyed by the early successes of Mercury—was beginning to wane, and by 1963, with reports of problems and possible delays, criticism came from several fronts. Some in Congress felt Kennedy should be spending more money on strengthening military efforts rather than on the somewhat intangible, far-off missions to the Moon. Even some leading scientists suggested that Project Apollo was a distortion of national priorities and many worthier pursuits existed for the funds being spent on landing humans on the lunar surface.

All this led to tensions among NASA leaders on how to proceed. It was no secret a strained relationship existed between administrator Webb and deputy administrator of manned spaceflight Brainerd Holmes. As stresses heightened, Holmes submitted his resignation on June 12, 1963. It took NASA more than a month to find a replacement, George Mueller (he pronounced it like "Miller"), who managed the Minuteman ICBM program at a space technology research laboratory. He reported to his new job on September 1.

If everyone at NASA thought Holmes moved as a whirlwind to get things accomplished for Apollo, Mueller must have looked like a hurricane. He covered vast swaths with decisions across NASA and was indefatigable, traveling to all the NASA centers frequently. The day of the week and time of day meant nothing to him. He scheduled meetings on Saturdays and Sundays and in the evenings.

Dr. George Mueller being sworn in as associate administrator for the Office of Manned Space Flight for NASA, by Dr. Hugh L. Dryden, NASA's deputy administrator. The ceremony took place at NASA HQ in Washington, DC, on September 3, 1963. Credit: NASA.

Mueller was used to applying a systems-engineering approach, where he could look at how all the parts fit together to create an interrelated whole. He felt NASA needed the same cohesive tactics. Soon after he arrived at NASA Headquarters in Washington, DC, Mueller focused his attention on relationships between his office and all the NASA centers, as well as all the contractors. He implemented a major reorganization, in which the heads of the field centers working on Apollo reported directly to him. Mueller made known he was the undisputed boss: the chief executive officer, chairman of the board. There would be frequent top-level meetings to discuss any problems and decide how to go about solving them.

Realizing NASA was understaffed in many areas, he also made a number of key personnel changes, bringing in new people and reassigning others. Reorganization at MSC aimed at strengthening the Apollo and Gemini program structures, where management was regrouped into seven major areas, and personnel who performed during the highly successful Mercury flight program were reassigned to support the upcoming missions. Mueller assigned George Low and Joe Shea—who both welcomed their new assignments—to Houston, Low to become deputy director under Robert Gilruth and Shea to head the Apollo Spacecraft Program Office. Air Force Brigadier General Samuel Phillips took over the Apollo Program Office at NASA Headquarters.

"George came in and did a remarkable job," said Shea. "One thing he did was to get more senior people running the program. We had a hard time hiring people from industry to come in and take jobs in NASA. Obviously, they were going to take a big reduction in pay. But he had a list of something like thirty Air Force generals and admirals that he wanted, and, by golly, we got most of them. That was something."

Mueller instigated an independent review of the agency's overall activities, and it quickly became apparent that Apollo would not make it to the Moon until well into 1971 without a more aggressive approach to its Saturn V rocket development. To deal with the schedule slips and budgetary issues, Mueller soon announced a new strategy to prepare for missions to the Moon known as "all-up testing" rather than testing components separately, as had been standard practice since the earliest days of NASA.

The idea was simple: For every launch, put the final version of the rocket on the pad with as much real flight hardware as possible, rather than testing each rocket stage bit by bit and each piece of the spacecraft separately. Ultimately, Mueller wanted the first flight of the Saturn V to be conducted with all three stages fully fueled, carrying a live Apollo Command and Service Modules as payload, with a sophisticated trajectory that would permit the CM to reenter the atmosphere under conditions simulating a return from the Moon.

Every test flight they could take out would save at least three months and lots of money.

So the first thing Mueller did was cancel the flights of the Saturn 1 booster—which were going to be used only for test flights—so that attention could be shifted to the upgraded Saturn 1B, which would use the same upper stage as the Saturn V. Mueller sent a teletype message to the Apollo field centers proposing the new accelerated schedule of Apollo flights, reiterating his "desire that 'all-up' spacecraft and launch vehicle flights be made as early as possible in the program."

"The first schedules had about ten incremental vehicle tests in them," said Shea. "We finally cut it down to two. In other words, on the first launch, all stages would be a real stage rather just putting ballast in the second or third stage. Put a lot of telemetry on it, and if it all works, by God, you're ahead of the game already. Then shoot a second one up to make sure the first one wasn't a random success."

Members of NASA leadership (L-R): Dr. Charles Berry, MSC medical director; Donald (Deke) Slayton, director of Flight Crew Operations; Eugene Kranz, flight director; Charles Matthews, Gemini Program manager; William Schneider, Gemini mission manager; General Leighton Davis; Dr. Robert Gilrath, MSC director; and George Mueller, associate administrator for Manned Space Flight. Credit: NASA.

The rocket engineers at Huntsville, in particular, were aghast. Von Braun and his team argued that by combining testing, it would be impossible to pinpoint where a particular failure occurred. The Saturn rockets were not being mass-produced like the Minuteman missiles Mueller previously dealt with. From von Braun's point of view, the two programs were unalike and could not be managed in the same way.

Mueller countered that sequential testing on a stage-by-stage basis just spread the risks among many tests rather than minimizing them. Mueller also knew that the Apollo program would not reach the goal set by President Kennedy using von Braun's approach.

Von Braun later wrote that "George Mueller visited Marshall and casually introduced us to his philosophy of 'all-up' testing."

Actually, it was more of an edict.

"Selling it to Wernher was not easy. He said it would never work, but we sold it to Wernher," said Shea. "We just finally told him that's the way it's going to be."

And it was.

THROUGHOUT THE SUMMER AND INTO the fall of 1963, John Painter and George Hondros worked to understand the Apollo Unified S-Band System and write their technical reports. One of the first things they did was contact Christopher Kraft's Flight Operations Division and have them compile a set of operational requirements for the USB system. Flight Director John Hodge was a British-born engineer who came to the Space Task Group from the Canadian Avro Arrow project and was one of the two flight directors for Cooper's Mercury 9 mission. Hodge agreed to meet with Painter and Hondros for an entire day—a Sunday no less—and by 5:00 that afternoon the three engineers had an approved set of requirements for the system.

Painter and The Greek had visited both North American and Grumman to understand how each company would incorporate the Apollo radio system into the CM and the LM. Grumman was still working on the design of the LM and hadn't yet finalized their requirements for the communications system. Grumman admitted they could use all the help they could get and were happy to work with Painter and Hondros to iron out the details.

North American, on the other hand, wasn't excited to have two outsiders from MSC come to Downey to tell them how to define the requirements for the radio system.

"To understand the magnitude of what we were trying to do," Painter said, "take a look at the complicated organization all of the contractors who were to provide the hardware making up the system. North American's subcontractor for communications was Collins Radio; however, Collins chose to build only part of the communications hardware, so they, in turn, took on a sub-subcontractor for the actual S-Band hardware, which was Motorola, Inc., of Scottsdale, Arizona. So just for the command module, we had three industrial companies involved in the design of the system."

Grumman's communications subcontractor was RCA, another major electronics company, headquartered in New York City. But, as with the CSM, RCA sub-subbed the actual S-Band hardware to Motorola.

There would also be a prime contractor and other subcontractors for the ground stations, which would be built in California, Spain and Australia.

"Now, since the entire system closely resembled a system JPL had invented, it was decided that at least part of the ground station hardware would be acquired through an extension of JPL's contract from their S-Band contractor," Painter said. "And guess who that was? Motorola."

Ah, finally some continuity in the whole process.

By this time, Goddard had set up a project office to start organizing the S-Band ground stations. The Greek started spending time in Maryland, helping them with the specifications, requirements and documents. In the meantime, George Mueller had asked for an investigation of the Apollo communications setup to see if everything was in order; Painter and Hondros worked with the investigator to provide MSC's plans and ideas on the USB. The work hours were piling up.

"Mind you, the system still hadn't been designed as a whole system—for both the spacecraft and the ground," Painter said, "but work was necessary now, because of the short schedules and long procurement times involved. Something had to be specified now."

Barry Graves organized a meeting to be held in November 1963 to get the MSC and Goddard people together to unify the plan and reach an agreement on how to proceed.

ON NOVEMBER 16, 1963, PRESIDENT KENNEDY
flew on Air Force One to Cape Canaveral, where he was met by Kurt Debus, Robert Seamans, James Webb and Florida senator George Smathers. They briefed Kennedy on the progress being made on Apollo and provided a helicopter tour to see the VAB, the new launchpads and other sites.

A few days later, Kennedy came once more to Houston. Although he was in town for a political fundraiser and didn't visit MSC, there was a motorcade through the city of Houston and the route would pass by some of the MSC temporary sites. NASA employees were encouraged to participate in the event.

Norman Chaffee brought his wife and young daughter to stand with several other MSC staffers and their families on Broadway and wave little US flags as President and Jackie Kennedy passed. Those who worked in the Houston Petroleum Center had only to step outside their building to see the motorcade and cheer on the Kennedys. The procession was a big event, with the Gulf Freeway closed in one direction so the presidential motorcade could pass by an estimated seventy-five thousand enthusiastic Houstonians. That night, Kennedy spoke of the future.

"In Texas and the nation . . . growth has meant new opportunities for this state, progress has meant new achievements," the president said, "and we dare not look back now. In 1990, the age of space will be entering its second phase, and our hopes in it to preserve the peace, to make sure that in this great new sea, as on Earth, the United States is second to none."

President Kennedy, center, attends a briefing at Cape Canaveral for an update on the Apollo program. At the far left sit NASA administrator James Webb and Vice President Lyndon Johnson. Credit: NASA.

The next morning, November 22, the Kennedys flew to Dallas.

That same morning, the first meeting was held between MSC and Goddard Space Flight Center to coordinate the ground communications with the Apollo USB on board the spacecraft. W. Paul Varson, who was leading the efforts at Goddard, and his technical assistant, Fritz von Bun, flew to Houston to join the meeting with several flight controllers, including John Hodge and Rod Rose—two members of the original Avro Arrow team—along with Howard Kyle, John Painter and George Hondros from the GSPO.

By noon, the group had made good progress and the administrators, flight controllers and engineers were comfortable with one another, having aired a number of management and technical issues to everyone's satisfaction. They adjourned and drove to a nearby restaurant for lunch. On the way back, Painter punched in the car's radio to get a little music. Instead, there was a special news program. Painter asked everyone to be quiet so he could hear better. Then came an announcement that President Kennedy had just been killed in Dallas.

"When we reassembled for the meeting, we just sat and looked at each other," Painter said. "Soon, someone suggested that we just go home and reassemble the next morning. That was what we did."

The news shocked the country and the world. Everyone found it hard to concentrate. People were glued to the radio and television, following everything that happened over the next few days: Lee Harvey Oswald was arrested and then was dead two days later, shot by Jack Ruby. The swearing in of a new president, the procession and funeral. A three-year-old boy saluting his father's casket.

A week after being sworn in, in the rear galley of Air Force One, the new president, Lyndon Johnson, announced in a nationwide television address that the launch center in Florida would be named in Kennedy's honor.

And work continued on Apollo.

"It was a shock here in Houston, just like it was all over the nation and the world," said Norman Chaffee. "But I can't say that Kennedy's death had an impact on what we were doing or that it made us rededicate ourselves, because we were already dedicated. We were already totally committed to getting to the Moon."

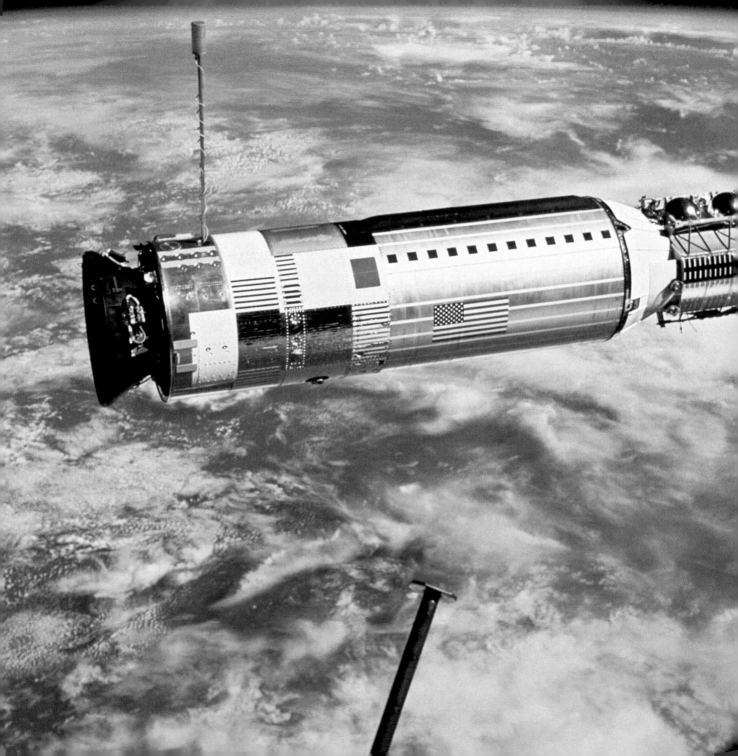

CHAPTER 3

1964

We figured it out as we went along.

−DICK KOOS, Simulation Supervisor

IN JANUARY 1964, THE US NAVY SHIP *Longview* slipped silently through the Pacific Ocean waves on a covert operation. Sailing 50 miles (80 km) off the coast of Hawaii, the converted World War II Victory ship was retrofitted with state-of-the-art telemetry antennas and communications transceivers. Its mission: locate a small capsule returning from space.

The *Longview* was part of a secret project, code-named Corona. This was the United States's first space program, a clandestine Cold War intelligence project. It was operated jointly by the CIA and the US Air Force, with the goal of acquiring the first-ever satellite reconnaissance imagery of the Soviet Union and China.

Corona started in 1959, in the days when US rockets exploded or crashed more often than not. It took fourteen launches to finally get a mission in orbit where all the components worked right, and it wasn't until 1964 that operations became routinely successful. But it would go on to be a remarkable program, with 144 launches in a dozen years.

The USS Longview, *sailing near Hawaii. Credit: Bill Wood, Operations Supervisor, Bendix Range Systems Department, Air Force Western Test Range.*

Left: An Agena Target Docking Vehicle in Earth orbit, during the Gemini 12 mission. Credit: NASA.

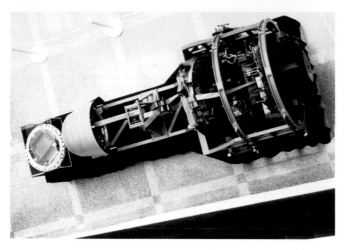

The Corona camera system. Credit: National Reconnaissance Office.

An image from Project Corona showing a missile site in the Soviet Union. Credit: National Reconnaissance Office.

The mission plan for this secret space program was audacious: launch a high-resolution camera to low Earth orbit and snap pictures from 100 miles (161 km) up, then eject the film canister to plummet from space, down through Earth's atmosphere where an airplane would snatch it in midair. Sounds crazy, but it worked. Corona's cameras acquired photographs on traditional film (digital photography wouldn't be invented for another dozen years or so) and stored the exposed film in an onboard capsule. Operators at a ground station would command the spacecraft to eject the film capsule, and the Air Force would deploy recovery ships and aircraft over the Pacific Ocean.

The trick was determining the falling capsule's trajectory.

"On the recovery tracking ships for Corona we would go out and sit underneath the path where the spacecraft would be going over," said Bill Wood, who oversaw a crew of about a dozen on the *Longview*. "And along with ground tracking stations on Tern Island and Kaena Point on the western tip of Oahu, we'd track the orbit of the Corona spacecraft and monitor the release of the reentry package into the atmosphere. Between all of us, we could pinpoint exactly where the capsule was coming down."

The crews learned to track the trajectory of the small spacecraft with such accuracy, they could send precise coordinates to pilots on board Air Force C-119 or C-130 aircraft. The pilots could then snag the parachutes of the film capsule with a special snare on the back of the planes.

"If necessary, we could deploy helicopters and swimmers to pick it out of the ocean if the USAF aircraft missed it on the way down," Wood said.

The CIA's cover story was that Wood and all his colleagues were working on a project called Discoverer, described publicly as a "scientific space program with a focus on biological research." The Discoverer program did actually employ scientists who developed habitable space compartments for mice and monkeys, and several of these compartments were launched during the early Corona flights, sometimes in combination with the camera and sometimes alone. In a couple of the early failures, the biological containers crash-landed and were recovered by people in different countries—one ended up in South America, and another compartment was possibly recovered by the Soviet Union. This made headlines at the time (inspiring the novel and subsequent film *Ice Station Zebra*) and, somewhat by coincidence, fell right into the cover story that these returning space capsules contained animals, not reconnaissance film.

Besides reaching its goal of gathering photographic surveillance, Corona made remarkable advances, ultimately helping NASA in many aspects. Corona proved that an object could come back from orbit and be recovered (an unknown when the secret program began), it paved the way for splashdown recovery for human missions, it demonstrated ground control and orbital operations and it proved it was possible to track a small object with incredible accuracy. Enough accuracy, in fact, to be able to send humans to the Moon.

Also, because of Corona, the CIA compiled highly technical data on Soviet space operations, keeping NASA in the know about the USSR's upcoming launches and buildup of launch facilities.

"A small group of us at NASA were cleared for the overhead photography that was available," said Robert Seamans. "It was very closely held at that time. But we had one room at NASA Headquarters that was built so that we could actually have somebody come over, a briefcase chained to their wrist, and they'd take out pictures and show us, then rechain it to their wrist and leave—all that kind of business. It was a glass room inside of a room, so that there was no window to the outside for people to get the vibration of the window and intercept the conversation that way, and all this spook stuff."

NASA received regular briefings from the CIA, so the space agency knew ahead of time about many of the Russian launches. NASA found they could ascertain the size of the rockets from these images because instead of taking rockets out to the pad vertically, the Soviets rolled them out horizontally on a railroad track.

In 1964, images showed a new massive launchpad under construction, confirming the circulating rumors about a large launch vehicle designed by Russian rocket engineer Sergei Korolev, purportedly called N-1, capable of launching humans to the Moon or beyond. With NASA's overarching goal of being the first to land humans on the Moon, knowing what the Russian space agency was up to encouraged NASA that maybe, finally they could get one step ahead. The CIA assured NASA that the Soviets didn't haven't any surveillance capabilities like Corona; but of course, there was no way to know for sure.

The intelligence gathered by Corona's cameras also showed the USSR's buildup of planes, intercontinental ballistic missiles (ICBMs) and other arsenals were not as extensive as that country claimed. Consequently, the US felt they didn't need to build up their own arsenal at breakneck speed, which—on the US side at least—helped cool tensions. Because of this reconnaissance, Corona would in due time profoundly alter the course of the Cold War. In April 1964, the US and USSR simultaneously announced plans to cut back production of materials for making nuclear weapons.

AGENA PROJECT

AGENA . . . versatile, upper-stage rocket vehicle employs a single rocket engine which provides 16,000 pounds of thrust. The engine can be shut down and re-started in flight through ground command signals.

Agena and its payload ride into space aboard a large booster rocket. Following staging, the Agena engine "first-burn" maneuvers the vehicle and its payload into an earth-oriented parking orbit. The Agena "second-burn" is geared to each particular mission - for example, an elliptical earth orbit or the ejection of a payload on a trajectory to the moon or planets.

This photographic exhibit presents the Agena missions . . . managed by the Lewis Research Center since January 1963.

National Aeronautics and Space Administration
Lewis Research Center

Agena project overview. Credit: NASA/Glenn Research Center.

Corona provided one more thing to NASA: a highly reliable vehicle for the upcoming Gemini rendezvous demonstration missions. Corona launched on a Thor rocket with an Agena second stage. As the Agena's capabilities developed through Corona launches, it became a dependable workhorse, just the kind of vehicle NASA needed for testing rendezvous and docking in space. Agena was the first space vehicle capable of multiple restarts during flight, and its navigation system could handle commands from the astronauts or ground stations.

"I worked in real-time flight operations for the Agena at the USAF Satellite Test Center in California right after I got out of the Air Force in 1960," said Gerry Griffin. "We launched some of the first spy satellites out of Vandenberg Air Force Base about as fast as they could pump the Agenas out of Lockheed's factory at Sunnyvale. In the early days, we were putting about as many satellites into the ocean as into orbit. But we figured it out, and I learned a lot—maybe most importantly, I finally learned what an orbit was and how you shape it!"

Flight Directors Gene Kranz, Glynn Lunney and Gerry Griffin. Credit: NASA.

But the entire time Griffin was at Lockheed (and when he later moved to General Dynamics), all he really wanted was to work at NASA. He watched the Mercury launches and kept tabs on NASA's every move; he just wanted to be part of that fray.

"I had actually tried to go to work for NASA in 1962, and the guy that interviewed me was named Gene Kranz," Griffin said. "And Kranz and I couldn't get together on money. And so I said, 'To heck with you,' and I went off and I did something else. But I knew I wanted to get into NASA so badly that, finally two years later, in 1964, I swallowed my pride and my wallet and took a pay cut to join Mission Control in Houston. It was one of the best decisions I've ever made."

Griffin wanted NASA because of the adventure and challenge, and NASA wanted Griffin because of his expertise with Agena, so he started out as an Agena flight controller. But about a month later, Kranz asked if Griffin would mind switching from Agena to become a flight controller for Gemini.

"There was actually a shortage of flight controllers," Griffin said. "When I got there, the unmanned Gemini test flights were just starting, and they needed some more expertise for Gemini. I think the fact that I did have some experience

because of being part of the real-time spaceflight operations for Corona was the reason Kranz decided to move me over into the Gemini side fairly quickly."

Kranz put Griffin to work as a flight controller responsible for the guidance, navigation and control systems on Gemini, call sign GNC. He loved it from day one.

As Griffin got to know his new bosses, he noted that both Christopher Kraft and Gene Kranz had an uncanny ability to find people who could respond to the pressure-cooker, split-second decision-making environment in Mission Control. Since NASA was basically inventing the process of flight control as the missions and programs progressed, there was no cadre of preexisting flight controllers or flight directors to choose from. And even though no college had curriculum for how to be a flight controller, MSC soon developed the capability to identify and hire young engineers and "home-grow" them to do the flight control job, mostly through on-the-job training. The rapidly developing technology of digital computers required a steep learning curve for essentially everyone in NASA. A few of the young engineers had been exposed to computers in college, but it was a new capability, and NASA was actually driving much of its expansion.

"Not only were we going to have computers in the spacecraft, we were going to have computers in Mission Control," said Griffin. "But not a lot of us had experience with computers. So, that's when Kraft and Kranz and others at MSC started finding people from colleges, universities and within the military that had some type of computer experience or aeronautical flight control experience. It was a matter of using all our country's assets to get to the Moon."

NASA needed expertise wherever they could find it, because several test flights of both the Apollo and Gemini systems were scheduled for 1964. First came a successful test of the Saturn SA-5 on January 29, as a nationwide television audience viewed the launch of "the most powerful space vehicle in the free world," as reporters called it. (SA-5 was the first Saturn rocket to fly with two stages.) Then in May, a Saturn 1 launched the first boilerplate version of Apollo Command Module (CM), a nonfunctional CM with the same mass to test the Saturn Apollo launch configuration.

Sandwiched between January's and May's launches was the first Gemini test flight in April to verify the structural integrity of both the Titan launch vehicle and the spacecraft, which successfully reached orbit. Then in September was another Saturn 1 test flight, providing final verification of the Saturn 1 rocket and successfully testing the Apollo CM's reentry capabilities.

These test flights helped NASA make significant strides toward the upcoming missions. Everyone at NASA, especially the astronauts who would fly these missions and the flight control team who would guide them, was encouraged that so many milestones had been reached. But where the flight controllers and astronauts would really learn their trade would be in the flight training simulations. Or as astronaut Mike Collins would later say, simulations became the "heart and soul of NASA."

OVER A BEER AT THE MOUSETRAP STEAKHOUSE

in Cocoa Beach, Florida, Harold Miller and Dick Koos made a major decision.

"We said, 'Gosh, we're going to have to use digital computers in the flight simulations for Gemini and Apollo,'" said Koos, one of the few members of the Simulation Task Group (or Sim Group) who had computer experience at that time. "The 'sims' for Mercury had been frustrating enough because all we had was an analog computer. And we knew we weren't going to do a very good job going forward without involving better computers. But we also knew it was going to be a challenge."

Harold Miller, left, in a discussion with NASA's public affairs officer John "Shorty" Powers, astronaut Alan Shepard (back to the camera) and Art Hand in the Mercury simulation area in the Mercury Mission Control building. Image courtesy of Harold Miller.

For the Mercury program, the big challenge for Miller, Koos and their small Sim Group had been unriddling the complex problems of not only doing flight simulations for the astronauts but also adding training for the fledgling flight control team. Just figuring out how to pretend you were flying in space was difficult enough, especially when no one had been to space yet.

"In planning for the first space flights, Chris Kraft and a guy named Jack Cohen had the idea to combine the astronaut crew training together with the flight controller training," said Koos, "but there was some disagreement about doing that, because some people just wanted to focus on the astronaut training. But Kraft and Cohen won out. But of course, nothing like that had been done before, so we just figured it out as we went along."

Their small group included Glynn Lunney, Stanley Faber, Miller, Koos and a few others, and together they organized the concept of what came to be known as integrated simulations. Kraft and Cohen's idea was this: Since the astronauts and flight controllers needed to work together during the missions to figure out how to deal with problems, why not have them train together so they could practice doing just that? In the early days, the Sim Group figured out how to tie in the displays and switches on a Mercury capsule cockpit trainer (which McDonnell Aircraft had built for the astronauts) with the console displays at the Mercury Control Center (MCC) at Cape Canaveral. The Sim Group would conduct practice runs for every mission, inserting all the problems they could think of, preparing the flight crew and the flight controllers together for all the possible contingencies in all the various phases of flight.

The Mercury simulation area at the Mercury Control Center at Cape Canaveral. Image courtesy of Harold Miller.

"In Mercury, the launch sims were pretty easy because there were only a couple of things we could do—either an abort or a nominal lift-off," said Koos. "During the simulated flight was where we started exercising all the different problems we could."

They found ways to simulate the loss of cabin pressure and other gauge indications to induce the call for an abort or flight modification. The temperature data had to be faked, and medical problems were faked with either the flight crew or controllers.

"We could at any time walk down to a controller and tell him that he was experiencing a heart attack or some other debilitating condition that would require him to leave his position," said Miller, who would later become head of the Simulation Branch at the Manned Spacecraft Center (MSC), "and the flight director would have to deal with the lack of a controller. Serious on-orbit medical problems were not simulated because we had a sim rule on not giving the controllers a problem that couldn't be solved or was deemed unrealistic."

One time they simulated the loss of the entire MCC.

"We shut down all the power in the building, and the flight controllers considered this an unrealistic failure," said Miller. A few weeks later, a bulldozer cut through the power cables leading to the building, rendering the MCC dead for hours. Fortunately, this occurred several weeks before one of the actual missions. The flight control team quickly developed backup procedures.

But that was the goal, to get everyone to think about all what-ifs—all the possible things that could go wrong so they could develop a plan and have solutions at their fingertips.

"Because of our sims, the flight controllers were developing these things called Mission Rules," Miller said, "where if a certain thing happens, you do such and such or if this other thing happens you do something else. They went through each of the systems and decided what was really critical, and if you lost it, would you have to abort, or could you do a workaround?"

The sims allowed the flight directors to hone their skills and instincts in leading their teams. Gene Kranz became a master at coming up with contingency plans, while Kraft prodded his controllers to think of every possibility.

Over time, the Sim Group developed cohesive procedures for how they conducted the simulations and they, too, came up with their own set of rules for sims:

- Simulations were to exercise procedures, interfaces (both human and hardware), Mission Rules and so on. They were not to teach systems. Classroom training was handled separately.

- The simulations were used to screen the flight controllers, to weed out people who were not adept at real-time operation.

- No grades were ever given or implied. It was the job of the flight director to decide on the readiness of the team and to choose the flight controllers.

- And of course, no catastrophic failures were allowed. There had to be a way out.

"The key decision, and I think this was critical," said Miller, "was how much to involve the Control Center technical staff, the backroom guys, who were the experts on each system. The decision was to keep the interface as simple as possible and to involve as many people as possible in the simulations."

But just as the Mercury flights got under way, Kennedy's challenge to land on the Moon came along. As details of NASA's future goals emerged, the Sim Group knew they'd be tasked with an even bigger job of simulating complicated rendezvous missions for Gemini and even more complex missions going to the Moon with two spacecraft for Apollo.

Before Mercury was over, Miller left Cape Canaveral for Houston to formulate plans for the Gemini and Apollo sims. First, he needed to define the interfaces between the Simulation Control Room and the two Mission Operations and Control Rooms—known later as Mission Control—under construction in Building 30 in Houston. Miller and his team had input on where the sim control room should be located and how it would operate. The decision to have two Mission Control Rooms for redundancy in case of an emergency also allowed for sims to take place in one room while an actual mission was operated in the other. Two Mission Control Rooms meant two Sim Control Rooms too.

"I decided to have a simulation control room located next to each Mission Control with a viewing window," said Miller. "This put the sim guys in close contact with the controllers and helped coordinate our activities, especially the debriefing after each of the sims. We could monitor and inject failures into both the ground and the spacecraft, to the devilment of the flight controllers and crew."

The crew cockpit trainers were going to be placed in Building 5, so all the wiring and hookups between the various simulation systems were housed in a series of underground tunnels at MSC. This sophisticated closed-loop simulation system was called the Simulation Checkout and Training System.

Once these decisions were made, all the contractors building the crew trainers, the control rooms and the consoles were able to develop the technical details of the many interfaces and the algorithms for simulating the environment, providing realistic data for the flight controllers and crew.

Transfer of IBM computers and equipment from the University of Houston to the Real Time Computer Complex at MSC. Credit: NASA.

Gemini became the first spacecraft with a digital computer and digital telemetry. And the new Mission Control would be on a digital-based mainframe system. These digital capabilities lent themselves to digital-based simulation systems, providing the most realistic way of doing a "let's pretend we're in space"–type training. But designing the simulations themselves for both Gemini and Apollo at the same time presented a huge challenge to the Sim Group, as they had to incorporate the different computers that would be used in each of the spacecraft. These computers were being designed and built at the same time as the Sim Control Rooms and Mission Control. Modifications were frequent, meaning almost everything was a moving target.

While the simulations for each program would be different, both could take advantage of the new IBM 7094 computers being installed for the Real Time Computer Complex (RTCC), also in Building 30 on the first floor. These were the biggest and fastest computers available, but it quickly became apparent that the Sim Group would need their own computer solely to support what they needed to do with the simulators and the flight controllers. This computer became known as the Ground Support Simulation Computer (GSSC). The sim team used the GSSC to build math models of the rockets and to simulate the control center telemetry systems, along all the data for the tracking and communications systems. IBM also had the job of creating all the simulation software, built from a set of detailed requirements provided by the Sim Branch and the flight operations team.

"Simulating the dynamics of a vehicle, especially with developing the trajectory and doing it realistically put a big load on the computer," said Koos. "It got pretty complicated real fast."

Bigger spacecraft meant an exponential explosion in the amount of data that needed to be processed. For Mercury, there were only about eighty-eight parameters that had to be transferred between the spacecraft and the flight control team. For Gemini there were about two hundred, and Apollo had close to fifteen hundred instrument data, such as temperature and pressure sensors on every tank, thermocouples buried in the ablative heat shield and astronaut health information. The simulators needed to be able to display the same type of data.

"The Gemini and Apollo flight crew trainers were so much more complex than the ones for Mercury," said Miller, "and the foremost challenge for Gemini was the addition of the Agena, which was a highly classified military vehicle, so keeping the system current presented real issues. We asked IBM if they could do a digital simulation of the Agena, which was a major design challenge since at that time there had never been a completely digital simulation. But it was the only way I could see us ever keeping a configuration similar to what the flight controllers would be seeing in the real missions."

Among Miller's many tasks was to hire more people in order to meet the demands of figuring out these complicated simulations, which was the only way NASA could meet the upcoming flight schedule. The new hires had to first learn what spaceflight simulation was and then work out all the details of how to do it.

"We were inventing the job at the same time we were learning how to do it," said Carl Shelley, who arrived in Houston in 1964. "And everyone had to figure out what to do from scratch, so we went ahead and did what made sense. We were all pretty young, and it was a fun time because there was nothing bureaucratic about anything. If you needed to do something, you just went and did it."

A view of the Real Time Computer Complex in Building 30 at the Manned Spacecraft Center in 1966. Credit: NASA.

As in almost everything NASA was doing at the time, this level of complexity of training through simulators had never been done before. There were snags, hiccups, troubles in developing the systems and a lot of long hours trying to get the system up and running (then to keep it running). In the meantime, the Sim Group had to develop plans for how to do the simulations, and there were also classes and training for both the flight controllers and the sim controllers in order to learn the systems and hardware inside and out.

"When I got here," Shelley said, "the flight controllers were developing their own handbooks, doing their own drawings from studying the manufacturing schematics and overlaying the mission data information to get the functional flow down into a minimum-size book they could refer to. And we in the Simulation Branch had to do the same thing, plus develop the malfunction procedures, which was pretty much anything a guy could think of that could go wrong."

Harold Miller in 1966. Credit: NASA.

The Simulation Control Area, with technicians on console. Back row (from left): Rod Rose (assistant to the flight director), Carl Shelley (SimSup), Harold Miller (Mission Simulation Branch chief), and Herman Mobley (tech support). Front row (from left): Gerald Griffith (Simulation Group), unrecognized person, Paul Joyce (Simulation systems engineer), unrecognized person and Bob Holkan (Simulation Dynamics specialist). Credit: NASA, caption courtesy of Carl Shelley.

They also had to set up a network sim of all the tracking stations around the world, which would still be used for Gemini, and for Apollo when the spacecraft were still in Earth orbit.

"We still had these remote sites all around the world, and we had to worry about how we would integrate those distant flight control teams into the overall training environment," said Shelley. "So we built a couple of simulated remote sites here at MSC, even though the guys were just through the wall in the next room. The computer would sequence them as if they were acquiring the vehicle from Bermuda or Australia or wherever during a real flight, turning the telemetry on and off into these simulated remote sites. It made the whole network look pretty real."

As the hardware for the simulators started to come together, the assignments were divvied up among the Simulation Branch: Shelley was in charge of sim operations, Koos oversaw all the systems and Gordon Ferguson was in charge of classroom training. Almost everyone took a turn at being a simulation supervisor, or SimSup.

Using this complicated, never-been-tried flight simulation system, the idea was to validate the plans and procedures for each flight and to make sure the flight controllers and crew would be able to handle any problems that might surface.

Would it work? The team had to wait until November 1964 when everything was built and ready. Astronauts Jim McDivitt and Ed White, command pilot and pilot for the Gemini–Titan 4 mission, were the first astronauts to be grilled with the new facilities.

"The first simulations run in Houston were Gemini launch aborts," said Miller. "I was acting as the SimSup and we ran nine cases on the first day. The most launch aborts we had ever run in a day at the Cape were four. It was quite a good feeling being able for the first time to run that many simulations with a brand-new complex."

AS THE INTEGRATED SIMS IN BUILDING

30 provided an intellectual challenge for everyone involved, NASA also developed other, more physical crew trainers and simulators for the astronauts, such as a centrifuge where astronauts were spun around to simulate the forces they would feel at launch, reentry and splashdown. The best way to simulate weightlessness on Earth came by flying a KC-135 aircraft in a series of roller coaster–like parabolas. The airplane would climb 10,000 feet (3,048 m) and then plunge downward an equal distance. The free fall simulated the feeling of weightlessness in 20- to 25-second intervals, during which the astronauts could practice various procedures, such as conducting tasks while wearing spacesuits. Some NASA training flights would do about one hundred of these climbs and dives. For obvious reasons, this aircraft earned the nickname the Vomit Comet.

By 1964, NASA was creating a couple of different trainers and simulators for landing the Lunar Module (LM), as well as mock-ups of lunar landscapes where the astronauts trained for their surface activities. At Ellington Air Force Base, the astronauts flew the Lunar Landing Training Vehicle (LLTV) to master the intricacies of landing on the Moon. This strange contraption looked so ungainly and awkward, the astronauts dubbed it "the flying bedstead." But it could accurately simulate the LM'S performance with LM-like thrusters for attitude control and allowed the astronauts to simulate flying in the Moon's one-sixth gravity. How it worked was all theoretical since no one had ever been to the Moon. Neil Armstrong called it "a contrary and risky machine, but a very useful one."

Indoors, NASA constructed special rendezvous and docking simulators for the astronauts to practice these maneuvers for both Gemini and Apollo, and new optics allowed for realistic out-the-window views of the starry views of space and the pockmarked lunar surface. Some of the high-fidelity cockpit trainers sat on a fixed base, others were on a moving base to simulate the actions of the vehicle. When the Gemini cockpit simulator arrived in Houston in April 1964, it was supposed to be stationed in Building 5, the Space Mission Simulation Facility. But that structure wasn't completed yet, so the cockpit trainer was put in Building 4, the Astronaut Office building. It barely fit, so getting in and out of it was

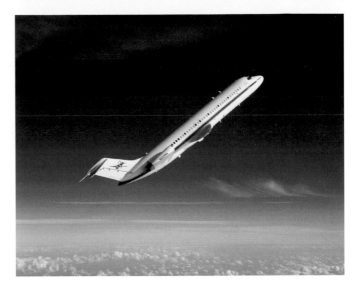

The "Vomit Comet," a NASA aircraft, flies a series of parabola patterns over the Gulf of Mexico to provide opportunities for astronauts and scientists to experience brief periods of weightlessness. Credit: NASA.

a challenge. With construction on Building 5 finally finished, Stanley Faber, head of the Simulation Branch, was anxious to get the simulator installed in its rightful place. But he got the runaround from the trades union about who was supposed to actually move it. Frustrated, Faber finally just decided to relocate it himself with a little help from his friends who worked in the Simulation Branch.

"My crew, which numbered something close to forty or fifty," said Faber, "picked up all the cables, took them across the street on their shoulders and spread them out on the floor in Building 5, and they rolled the cabinets across the street—they were all on wheels—over a Saturday and Sunday. On Monday morning when the union reps came storming in, we were finished and back in operation."

Besides being in trouble with the union and NASA management, Faber figures he probably broke a few child labor laws too.

"I had my two little boys crawling under the suspended floor, checking where wires went," Faber said. "We couldn't do it; we were too big. But they could fit under the floor. We'd hear a little voice, 'The wire's over here,' and we'd lift up that floor section and so we could get everything installed."

New astronaut class in 1964, "The Fourteen." Front row (from the left): Edwin E. Aldrin Jr., William A. Anders, Charles A. Bassett II, Alan L. Bean, Eugene A. Cernan and Roger B. Chaffee. Back row (from the left): Michael Collins, Walter Cunningham, Donn F. Eisele, Theodore C. Freeman, Richard F. Gordon Jr., Russell L. Scweickart, David R. Scott and Clifton C. Williams Jr. Credit: NASA.

NASA SELECTED A THIRD GROUP OF astronauts, who reported to Houston in the spring of 1964. The new group called themselves "The Fourteen"—bringing the total Astronaut Office headcount to thirty. The new hires were Buzz Aldrin, Bill Anders, Charlie Bassett, Alan Bean, Gene Cernan, Roger Chaffee, Mike Collins, Walt Cunningham, Donn Eisele, Ted Freeman, Dick Gordon, Rusty Schweickart, Dave Scott and Clifton Williams. The Fourteen weren't just military test pilots: Eight carried advanced degrees and two—Cunningham and Schweickart—were considered civilians, even though they had experience as military pilots. The group was selected from approximately 500 military applicants and 225 civilian applicants.

The astronauts arrived in Houston just in time to move into brand-new buildings at MSC. Construction of most of the facilities were finally completed, and from February through June 1964, more than twenty-five hundred employees relocated from their temporary offices and moved on-site at MSC.

"My first day in the office side of Building 30, there was stuff piled everywhere," said Shelley. "The hallways were littered with all the office supplies, from copying machines to papers. Everyone was just moving in and it seemed like there wasn't enough room."

Most of 1964 was chaotic until everyone got settled. With thousands of employees now arriving at one location to start the day, traffic jams and parking problems ensued, so staggered work hours were instituted, starting in May. But those who were already putting in long hours were seemingly unaffected by the change in work hours. For Miller and the Sim Group team, their typical day was twelve hours long, and they worked most weekends as well. Several Sim Group members and flight controllers commiserated about the fact Building 30 didn't have any windows, but it was just as well. They worked such long hours, they hardly saw any daylight anyway.

The world went on with most everyone at MSC oblivious to anything going on outside of NASA. Race riots gripped several cities when the Civil Rights Act of 1964 was signed into law. Boxer Cassius Clay became Muhammad Ali and the heavyweight champion of the world. President Lyndon B. Johnson escalated US involvement in the Vietnam War, and the Beatles took America by storm. Jay Honeycutt of the MSC simulation team would later say that he has absolutely no recollection of any music of the 1960s—he was just too busy to listen to the radio.

Aerial View of the Manned Spacecraft Center in 1964. Credit: NASA.

But one new facility at MSC allowed a brief respite from work, giving employees a chance to chat and catch up with their friends at work. The opening of the cafeteria—aptly named The Cafeteria—was a big deal, with a several-page spread of pictures and information about the menu written up in the *Space News Roundup* newspaper. The MSC staff greatly appreciated not needing to drive off-campus to eat lunch. The food was good and considered reasonably priced, with breakfasts ranging from fifteen to sixty-five cents and luncheons from fifty-five cents to a dollar. It also offered a place to get together for impromptu meetings or for a chance to see who might be at MSC that day. MSC started hosting several educational events for students and teachers, and The Cafeteria offered a good place for the students to perhaps catch a glimpse of an astronaut.

For the employees, though, the astronauts were just part of the workforce. Seeing them at The Cafeteria was a chance to ask about the latest change order for a specific component or get an update on when they would be available to take part in a systems test. Norman Chaffee ran into Roger Chaffee at The Cafeteria one day and introduced himself. The astronaut invited the engineer to join him for lunch and they discussed family trees. "We ended up having lunch together several times, often looking back at our lineage to see if we might be related. We never found a connection, but we did become good friends," said Norman Chaffee.

Classroom training for astronauts. Credit: NASA.

WITH THE NEW ASTRONAUTS' ARRIVAL,

John Painter and George Hondros prepared for another communications-training class, this time a two-week session, with two-hour classes each afternoon. The astronauts would also receive extensive instruction in rocket propulsion, aerodynamics, astronomy, physics, environmental control systems, survival and rendezvous and docking techniques. The primary objective of all training was to teach the astronaut trainees about their spacecraft and to familiarize them with flight conditions like acceleration, noise, heat, vibrations and disorientation.

At the first meeting of the communications class, Painter sized up the new group and quickly found they all had a sense of humor.

"I passed a blank sheet of notebook paper around, ostensibly to take a class roll, although I knew their names," Painter said. "What I really wanted was their autographs, but when I read the names, I laughed like crazy. Included were Bolivar P. Shagnasty, J. P. Four [a jet fuel] and Flash Gordon." Unfortunately, the signed sheet flew out an open window of Painter's 1955 Ford a couple days later.

The new astronauts kept Painter pressed up against the blackboard with their questions. He was especially impressed with Roger Chaffee, who seemed to quickly grasp communications theory and kept Painter on his toes. Rusty Schweickart was the most personable of the group and just as technically sharp. The third astronaut who drew Painter's attention was a quiet, slightly balding guy, who didn't say much. But Painter knew Buzz Aldrin had done his PhD work in the theory of spacecraft rendezvous, and they chatted briefly about the concept. The most notable aspect of the group, in Painter's mind, was that no one in this class went to sleep.

Also during that week, Painter and The Greek completed the manuscript of volume 1 of their "NASA Technical Notes" on the Unified S-Band communication system. After conferring with the astronauts, Painter and Hondros decided to utilize the information in the technical report for the second week of class. All that needed to be done was for the eighty-eight-page manuscript to be typed in final form. So at about 5:00 p.m. that Friday afternoon, the two engineers broke the news to their faithful secretary, Frances Smith, that they needed twenty copies typed and xeroxed by 2:00 p.m. Monday afternoon.

"Of course, she did it," Painter said. "Even though she tried to be mad at us, she just couldn't be. I'm pretty sure she worked about twelve hours that Saturday to type it up for us and make all the copies. She was something else."

PAINTER AND HONDROS CONTINUED TRYING

to ensure functional compatibility between all the different contractors for the USB radio system for Apollo communications. During a couple of meetings in early 1964 with all of the various contractors, the problems weren't resolved to anyone's satisfaction. Barry Graves Jr. expressed considerable concern, and early in 1964, he started making trips to Downey, California, to check out, among other things, how North American Aviation was spending their money allocated for communications.

But North American had bigger concerns than just the USB system. They were in the middle of redesigning the entire Command and Service Modules (CSM). When NASA awarded the Apollo contract to North American in 1961, the initial designs were based on the Direct Ascent plan for getting to the Moon. Therefore, their early version of the CSM, called Block I, was crafted to land on the Moon atop a rocket stage and it did not have a docking port for an LM. But with the change to the Lunar Orbit Rendezvous (LOR) plan, a substantial redesign was required and the CSM also needed to lose some weight. North American was also haggling with NASA over certain design aspects: They wanted to use a mixed-gas atmosphere inside the CM instead of pure oxygen and include a different type of hatch with explosive bolts for quick release in case of an emergency. NASA turned down both due to time, cost and weight (plus they didn't want a repeat of what happened with Gus Grissom's Mercury spacecraft, which sunk in the Atlantic Ocean after the hatch blew open unexpectedly following splashdown). Other technical obstacles also surfaced in several subsystems, such as environmental control and communications. Since a few of the Block I CSMs were already in production, NASA decided the most efficient way to keep the program on track was use the Block I versions for early Earth orbit test flights. The new design, Block II, would include a docking port and hatch and incorporate weight reduction and lessons learned from Block I.

But as far as Graves, Painter and Hondros were concerned, the communications-systems issues couldn't wait, and North American appeared to be ignoring them, not willing to make some of the suggested fixes—especially for redundancy and flexibility of the system. What was needed was laboratory verification of the system components, hooking together the prototype hardware USB designs for the CSM, the LM and the ground stations.

"We didn't have any laboratory facilities in Houston for setting up a compatibility test," said Painter, "and we were just preparing to move to the permanent center facilities at MSC. It was clear that to set up a test rapidly would mean contracting it out, and it was clear that Motorola had a decided advantage in terms of experience and familiarity with the system since they were building the spacecraft transponders for both the LM and CSM and also building a good part of the ground station hardware."

In the interest of speed, Graves quickly approved getting the contract set up. However, soon after word got out about the contract, a strange edict came down from NASA Headquarters. Motorola could be selected to do the tests. But the tests themselves would be done at MSC, meaning a whole new laboratory would need to be built before the tests could be run.

Then something even stranger happened. Just a few weeks later, Painter and Hondros walked into their offices one Monday morning to find a green slip of paper on all the desks. It was a notice that the Ground Systems Project Office (GSPO) had been abolished, and they were all to be reassigned. Additionally, Graves had been taken out of the Apollo program and would be going back to the Langley Research Center, with his management responsibilities drastically curtailed.

Everyone was stunned. Gradually, Painter and Hondros pieced together what they thought had likely happened. They were certain the reason was office politics, both external and internal to MSC.

"Graves had many disagreements with North American on many things about the Command Module that he was attempting to get them to fix," said Painter. "One, of course, was the Block II Unified S-Band. But there were far worse problems with the Block I design, which needed to be fixed as a part of the Block II design. The prime contractor was dragging its feet and Graves had been pushing."

The Apollo Unified S-Band System. Credit: NASA.

For instance, Painter knew that with respect to the Unified S-Band system, Graves threatened North American with cutting its profits on the system by having it installed in the spacecraft at the Cape by its manufacturer, Motorola, as government-furnished equipment. Painter was sure that North American pressured NASA to eliminate the demands from Graves.

The internal aspect, in Painter's opinion, was that another assistant director at Houston wanted control of the construction of the Mission Control Center, which had been in the hands of the GSPO. Thus, when Graves drew fire externally from the Apollo prime contractor, other administrators went right along with firing him.

Painter and Hondros were reassigned to a new division and they decided to do two things.

"First, to get volume 2, the mathematical analysis of the Apollo USB system, written, come hell or high water," said Painter. "Second, we decided to look for jobs away from MSC. We knew that it would only be a matter of time before an astronaut got killed in the Command Module, and we didn't want it to be on our watch."

NORMAN CHAFFEE, HENRY POHL AND
Chester Vaughan continued to live under the premise that there was nothing more interesting to work on than rocket engines.

"You get to do things that nobody has ever thought about and nobody else has ever dealt with," Chaffee said. "And then you also have problems like nobody else too."

An Apollo Command Module reaction control 93-pound (42-kg) thrust engine (left), and a cutaway showing the charred area of the thick wall that results from firing the engine. Credit: NASA/Mike Salinas and Norman Chaffee.

The problem under consideration by Chaffee and Pohl during most of 1964 came from cracked liners on the two sets of small Gemini thrusters. The Reaction Control System (RCS) controlled the reentry portion of the spacecraft, and the orbital attitude maneuvering system provided steering capabilities while in orbit. Both systems were embedded in the outer hull of Gemini, and under test conditions, the liners in the "throat" area of the thruster were cracking and sometimes flying apart into little pieces. If Chaffee and his colleagues in the Propulsion and Power Division at MSC couldn't figure out how to prevent the cracking and shattering, the Gemini spacecraft wouldn't be able to fly. If Gemini couldn't fly, neither could Apollo.

So this little problem became a big concern for both the Gemini and Apollo program offices, because the Apollo CM also contained these embedded small rocket engines. The thrusters were built by Rocketdyne, but Chaffee and his compatriots were test-firing them at their newly built test facility at MSC, the Thermochemical Test Area, where they could conduct qualification tests.

"We were fortunate to have our own independent design, manufacturing and test capability in Houston," said Pohl, "and we could evaluate the new designs Rocketdyne was proposing, as well as evaluate our own ideas."

"What was happening was something called thermal shock," explained Chaffee, "where if your thruster hasn't operated for a while, it cools down. We were testing the thrusters under flight conditions, and things get pretty cold in space. Suddenly

you turn on the thruster, fire it for fifty milliseconds and sock it with a shot of fifty-five-hundred-degree gases—it's like throwing an ice cube into a glass of tepid water. It cracks or sometimes shatters because of the tremendous change in temperature."

Chaffee had already spent a good chunk of the past two years improving on some of the design parameters of these thrusters, called ablative thrusters. After several months of testing and redesigning, Rocketdyne made some improvements in the ceramic liners. These helped but didn't completely solve the cracking issue. However, their tests also showed that none of the new thrusters were failing because of the cracking, and the design requirements were being met.

"Our program manager was a practical individual," said Chaffee, "and he said, 'Look, I've got a certain amount of money and a certain amount of time, and I've got to get on with this. Okay is good enough for me. It doesn't have to be perfect, as long as it works.'"

So NASA and Rocketdyne decided they needed to live with the problem because time was running out before the Gemini flights were to begin. With certain restrictions, cracked throats were to be accepted.

"We never did completely solve the cracking problem," Chaffee said. "So we defined a specification that said the throat piece can't break into more than twenty-seven pieces, and none of the pieces can be ejected or come out. That ended up working."

"One of the things that really helped us a whole lot in the Apollo program is that we were able to get things tested early on," Pohl said, "even prototypes or things that we knew sometimes wouldn't work. But at least it gave us an idea of how to change something, how to modify something, how to do something different. We were always doing something. We didn't have time to just sit around."

ON SATURDAY, OCTOBER 31, JOHN PAINTER was at home, working in his yard in Pearland, when four T-38 jet aircraft—the type of planes the astronauts used for flight training—flew over at a low altitude. Because of the planes' low altitude and the fact they were following the Friends-wood-Pearland Highway, Painter knew something must be wrong. He switched on the radio and heard reports of an airplane crash at Ellington Air Force Base. Painter couldn't help it—he needed to know what happened, so he jumped in his car and drove to Ellington. There, between the Gulf Freeway and Old Galveston Road, was the crash site, cordoned

John Painter in 2010. Image courtesy of John Painter.

Astronaut Theodore C. Freeman in 1964. Credit: NASA.

off. He found out that as astronaut Ted Freeman was making his landing approach to the Ellington runway, his T-38 struck a Canadian snow goose. The goose shattered Freeman's plexiglass canopy and the pieces went straight into the engine intakes, flaming them out. Freeman ejected from the aircraft but did not have sufficient altitude for his parachute to open. Freeman was found dead, still in his ejector seat. MSC mourned the first loss of an astronaut and the tragedy of an accident that occurred not in space, but during a routine aircraft flight. Freeman became the first American astronaut to lose his life in the quest for the Moon.

Painter recalled that he and Hondros had spoken to Freeman in passing just the previous afternoon outside the MSC headquarters building. Freeman's tragic death, along with the unsettled and challenging nature of Painter's work at that time, made it difficult to concentrate and he was secretly looking at different employment options. And then Hondros received an offer to work on the Apollo ground station design and implementation at Goddard Space Flight Center in Maryland. He would travel to the various ground station sites around the world, troubleshooting problems and overseeing the installations, which gave him the opportunity to visit his family in Greece occasionally.

"After George left, I was really lonesome and knew that my own days at Houston were numbered," said Painter. "George and I had been a good team, and we had done a lot of good for Apollo in a year and a half. But I just didn't feel like carrying on the fight by myself. Besides that, I was looking for more money, since I now had a wife and three kids to feed."

Painter made a few trips to Maryland to work with Hondros on the second volume of their technical paper. Once it was finalized and submitted for publication, Painter accepted a job with Motorola and moved his family to Scottsdale. He couldn't work on the Apollo USB system because of federal regulations concerning conflict of interest in the hiring of former government employees. Instead, he worked on the Air Force's version of the Unified S-Band system called Space Ground Link.

EVEN THOUGH NO ASTRONAUTS FLEW TO

space in 1964, this was a pivotal year for NASA. Preparing for the Gemini program defined and tested the technology and skills NASA would need to land on the Moon. Going forward, Gemini needed to test the ability to fly long-duration missions—up to two weeks in space—and understand how spacecraft could rendezvous and dock in orbit around the Earth and the Moon. Astronauts would need to leave their spacecraft for their extravehicular activities (EVAs), such as conducting spacewalks to demonstrate they could work effectively in bulky spacesuits in the harsh environment of space. These were all things that had never done before, and 1965 and Gemini would prove NASA could do them all.

CHAPTER 4

1965

Just [say] the magic words: "I'm working on Apollo."

—EARLE KYLE, aerospace engineer, Honeywell

JERRY BELL ISN'T SURE WHEN HE FOUND time to ask his girlfriend to marry him, but somehow during 1965, they tied the knot. This year turned out to be one of the busiest of his life: taking care of his recently widowed mother, courting his sweetheart and many times working sixty to seventy hours a week with the Rendezvous Analysis Branch at the Manned Spacecraft Center (MSC), developing rendezvous techniques for Gemini and for subsequent use with Apollo.

"The hours were just unbelievable," Bell said. "I had to get permission to take time off to go to my engagement party, which was on a Sunday afternoon. And the only time it worked to get married was on Thanksgiving Day."

But as 1965 began, so too began a flurry of activity, with the Gemini flights directly ahead and Apollo soon to follow. For Bell and his coworkers absorbed with the intricacies of rendezvous and docking, success for Gemini seemed almost mandatory.

"We worked in parallel on two programs at once," said Bell. "The whole purpose of Gemini was to verify and test out all the different concepts for Apollo. But there were certain aspects

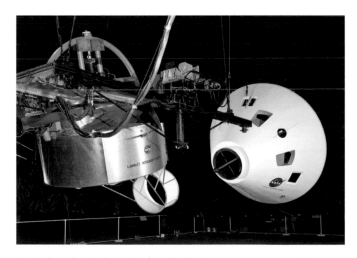

A simulator for Apollo rendezvous for the Command Module and the Lunar Module at Langley Research Center. Credit: NASA.

of Apollo that were completely divorced from anything we were doing in Gemini. And in the meantime, everything we learned from one Gemini flight went right to the next one."

Left: The Gemini 2 spacecraft in Earth orbit is connected by a tether to an Agena Target Docking Vehicle in September 1966. Credit: NASA.

Astronaut Edwin E. "Buzz" Aldrin, Jr. Credit: NASA.

Bell joined the rendezvous team in 1963, working with Ken Young, Catherine Osgood and an ever-growing group of engineers and scientists. Engineer James (David) Alexander also came on board in 1963, and the rendezvous team operated under the leadership of Ed Lineberry, who took over when Bill Tindall became involved with his wide-ranging responsibilities in coordinating the Data Priority meetings. Lineberry was a talented but shy mathematician and an orbital mechanics genius. He came up with the logic and equations for what he called the analytic ephemeris generator (AEG), a computer program that became an essential rendezvous-planning tool for the group. This allowed them to plan out different maneuvers for various rendezvous scenarios.

"At first we thought, *Just launch directly into orbit and rendezvous on the first orbit*," Osgood explained, "*but oh, if your launch is delayed, your plan is gone.* So then we looked at the tangential method, but then it turned out the closing rate then was just so fast that it would be a dangerous final approach. So then we came up with the concentric two-maneuver sequence, and following that we came up with what was called the CSI/CDH. Then, it was the NSR, the NCC/NSR burn, and . . . "

All of that may *sound* simple (to rendezvous engineers, anyway), but it actually took years to figure out the different components within rendezvous. Fortunately for the team, Buzz Aldrin came aboard. Among the third group of astronauts, he graduated from Massachusetts Institute of Technology (MIT) with a doctorate of science in astronautics, and his thesis was titled "Manned Orbital Rendezvous." Tindall easily convinced Aldrin to confer with the rendezvous team in 1964, and the astronaut helped develop the plans for how two spacecraft could meet in orbit and at the Moon. His work earned him the nickname "Dr. Rendezvous" from fellow astronauts, a named bestowed, Aldrin noted, with a mixture of respect and sarcasm.

"Buzz came over to our building to discuss exactly how we wanted to do the rendezvous coming up off the Moon," said Alexander. "We were throwing ideas around, and we found he was an absolute genius about rendezvous. Many of his ideas and suggestions were things we actually ended up doing, such as the co-elliptic sequence, which places the chaser vehicle on an intercept trajectory with the target vehicle."

Alexander himself designed several of the rendezvous maneuvers used throughout Gemini and Apollo, one called the conic fit, and another commonly referred as the football rendezvous because the motion of the Lunar Module's (LM's) trajectory looked much like the shape of a football.

For all the missions, the team needed to compute both nominal (normal) and contingency backup plans for all the things that might go awry.

"Frankly, for Apollo, we spent 80 percent of our time on the what-if contingencies," said Young, "like having the Command Module perform a rescue if the Lunar Module couldn't get back from low lunar orbit, which involved a long, complicated sequence that would take hours. The truth about mission planning is, our work involved guarding against and preparing for those one-off, weird, off-nominal events."

The Rendezvous Analysis Branch was organized under MSC's Mission Planning and Analysis Division. Since the Mission Planning and Analysis Division encompassed so many areas of study and work, the Rendezvous group found themselves working with several other groups and divisions at MSC. For example, they needed to coordinate with Henry Pohl and the engineering directorate to understand the Reaction Control System (RCS) thruster system in order to incorporate the thruster capabilities into the rendezvous maneuvers.

Engineers at the Manned Spacecraft Center. Credit: NASA.

Additionally—and surprisingly to some—computers and software became a large part of their work, mainly due to the need to combine the spacecraft hardware with the software for the Apollo Guidance and Navigation Computer.

"I really felt the software work we did was as important as anything that I worked with," said Osgood, "not in doing the coding myself or the equations, but in writing requirements for what needed to be done and then working with the program to look for any bugs in it."

The Rendezvous team added several different programmers to their team, and software specialist Bill Reini gained notoriety in the rendezvous world for coding a project that came to be known as "The Monster" because it could encompass several different rendezvous methods and numerous other concepts for spaceflight.

"Bill was sort of a reticent person," Osgood said, "and he'd never write anything down if he could help it, so I wrote his users' manual for his Monster. Not very many people could run The Monster, other than me, because it was really a chore. But, you know, I'd gotten accustomed to it and had a really good working relationship with Bill."

With the programmers on board, they could create programs and software—as rudimentary as they were in 1965—and get it done almost immediately.

"We now didn't have to wait for years and years for somebody to decide that they were going to work on that piece of software," Osgood said. "And if you'd find problems, you'd take it to the programmer, he'd grumble a little bit, and a few hours later he'd throw some cards on your desk and say, 'Here. Try this.'"

By *cards*, Osgood was referring to computer punch cards, the primary method of inputting data to a computer in those days. The digital data represented on the cards came in the presence or absence of holes, lined up in different columns, which the computer could read as ones and zeros. Once a card had been punched and completed, it technically "stored" that information. Groups or "decks" of cards formed programs and collections of data, but they needed to remain in order.

"We actually had a group of people that would keypunch the cards," Osgood said, "but we were usually a little too impatient, and there was a keypunch machine right here. We knew exactly where we wanted the information to be punched on the card, so we'd have big stacks of IBM cards, take them over to Building 12 and submit them to be run on the mainframe, the IBM 7094."

The next day, Osgood would get the cards and a printout back, but she could only submit about one run of cards a day, however. So, any mistakes meant it would be a day or two until she could try again. Or if something needed to be changed or added, another day would go by. And if the stack of cards were dropped or somehow got out of order? Start over.

How did they relieve stress?

"We would periodically take off half a day to go down to Galveston and go fishing," said Bell. "We had lives outside of work, but not much of it."

In the meantime, the Gemini flights started in earnest in 1965, becoming a link between Mercury and Apollo and paving the way for the ability to land humans on the Moon.

EARLE KYLE WAS RECRUITED FOR

another contract-engineering job for Apollo, this time in Minneapolis, his hometown. Honeywell received subcontracts from North American Aviation for several systems on Apollo, some involving analog systems and others using digital electronics, so Kyle's unique skills continued to be in demand. He worked at Honeywell for a year on a contract and then was offered full-time employment there. His family was thrilled to have the stability of not moving from place to place; Kyle was ecstatic because he could still work on Apollo.

"Sometimes I think I would have paid *them* to be able to work on this, no joke," Kyle said. "Holy crap, they were going to let me keep working on the spacecraft that were going to the Moon! It was like a dream come true."

Honeywell built a rich history of creating instrumentation for military aircraft as well as attitude-control systems and control columns or joysticks. Now, Honeywell and several other companies encompassed under a larger entity called Honeywell Aerospace contributed to the Apollo effort: developing the analog flight computer for the first stage of the Saturn V rocket to keep the entire rocket stack stable as it rose through the atmosphere, and building equipment for the Unified S-Band communications system on both the spacecraft and the ground stations. They also contributed to the Scientific Experiment Package, the instruments that would be placed on the Moon by the astronauts, which would run autonomously for long-term studies of the lunar environment.

Honeywell also developed the stabilization and control system that operated the thrusters on the CM and the LM and built an all-attitude display unit for the CM called the Flight Director Attitude Indicator (FDAI), commonly called the "eight ball" because it resembled the black eight ball in the game of pool. This device monitored the guidance and navigation system and provided a display of the spacecraft's orientation and attitude. The Honeywell plant in Minneapolis had been built well before World War II but just happened to be situated exactly on the forty-fifth parallel, halfway from the North Pole to the equator. This made it the perfect place to test gyroscopes and accelerometers for the space program.

Kyle worked specifically on specialized ground support equipment called the Bench Maintenance Equipment (BME) computer, which served to test hardware during production in Minneapolis and also acted as ground support equipment out on the launchpads at Cape Canaveral. He also enhanced the electronics for shoebox-size computers that would operate the thrusters and displays on the attitude-control system on the two spacecraft. He worked on refining the digital logic circuits, which figured out which small thruster engine to fire when a command was given either by the computer or by an astronaut using a joystick to control the spacecraft during the flight to and from the Moon.

GEMINI 3

The first crewed flight of Project Gemini launched on March 23, 1963, from Cape Canaveral in Florida, carrying Virgil (Gus) Grissom and John Young in the first US flight with two astronauts aboard. A Titan rocket boosted the small, cramped capsule to approximately 140 miles (225 km) above the Earth, and the crew made three orbits in the 4-hour-and-52-minute flight. The main objectives of the flight were to evaluate the two-man design of the Gemini capsule, test out an improved tracking network and orbital maneuvering system, and to attempt a precision-controlled reentry and landing while evaluating recovery procedures and systems. The mission was considered a success despite a few problems with the thrusters and the parachute, causing them to splash down over 100 miles (161 km) away from the intended landing site. Also during the flight, a surprise appearance of a corned beef sandwich—which Young had smuggled aboard—created a floating-breadcrumb mess.

Clockwise from top left: Launch of the Gemini 3 mission. The spacecraft was nicknamed "Molly Brown." Credit: NASA.

Gemini 3 astronauts Gus Grissom (left) and John Young. Credit: NASA.

John Young took this picture during the Gemini 3 mission as the spacecraft passed over Northern Mexico. The large light-brown area is the Sonoran Desert. The Colorado River runs from upper right to lower left. The lower portion of the picture is Mexico, the upper left is California and the upper right is Arizona. Credit: NASA.

Honeywell Bench Maintenance Equipment computer. Image courtesy of Earle Kyle.

It was a heady and hectic time. "Because of the time constraints of trying to get to the Moon before the end of the decade, we worked seven days a week, one-hundred-hour weeks sometimes," Kyle said. "I was never forced to do that, but if anyone had their heart in this work like I did, they worked as many hours too."

The gear from Honeywell controlled or interfaced with many major systems, including the large gimbaled engine on the rear of the Service Module (SM). This engine could be moved back and forth and side to side to alter the course of the ship on the way to the Moon and on the return to Earth. Just before the astronauts returned to Earth, they would jettison the SM so only the cone-shaped, heat shield–protected CM reentered Earth's atmosphere.

"The hardware I worked on helped make sure the entry angle was exactly −6 degrees below the horizon line—plus or minus 0.5 degrees," Kyle said. "Too low, and the CM could flip over, pointed end facing down, and would burn up. Too high and the CM could skip off the atmosphere and be lost to space. This is one of the most critical and dangerous parts of the mission and all three astronauts' lives were in our hands. That's why we worked so hard."

Every day Kyle would drive to the Honeywell plant, working some days until 2:00 or 3:00 a.m. On long days, he'd catch a nap somewhere within his five-person cubicle—on the floor or head-down in exhaustion on his desk—and then go back to work. If Kyle and his coworkers weren't solving problems on the design lab floor, they were working with subcontractors around the country who provided the nuts and bolts for all the equipment.

"Each of the supplier companies would have a crew available at all hours, especially if we were working on solving issues," Kyle said. "We found we could rule the world from our rotary-dial phones by just saying the magic words, 'I'm working on Apollo.' And before you could finish the sentence, you were connected to God in that company. They would bend over backward to instantly solve your problem, finding the bolt, diode, piece of glass or whatever you needed and then sending someone on a plane to bring a sack of parts to solve the problem."

During Apollo, an intricate link developed between all the smaller supply companies and Honeywell, and then Honeywell was linked to the other, larger companies and academic institutions that were contributing the expertise needed to reach the Moon.

An artist's concept depicting the Apollo Command Module, oriented in a blunt-end-forward attitude, re-entering the Earth's atmosphere after returning from a lunar landing mission. Credit: North American Rockwell/NASA.

"While the MIT Instrumentation Lab was writing the equations and software that would guide us to the Moon," Kyle said, "they had passed some of the hardware development of the inertial platform to the AC Spark Plug division of General Motors in Milwaukee, Wisconsin, and we were supplying a subset of components to them. So many weeks I was on the phone to people in Boston or Milwaukee or flying all over the country to some small company in places like San Diego where my expertise in vacuum-tube electronics really pulled the fat out of the fire for a weird failure problem on the big Bench Maintenance Equipment computer. Apollo was an amazing linkage of all these different companies and capabilities across the country."

FOR THE MIT INSTRUMENTATION LAB, one of the big worries about the Apollo Guidance Computer was reliability. The computer would be the brains of the spacecraft, but what if it failed? Since redundancy was a known solution to the basic reliability problem, Dick Battin and his colleagues suggested including two computers on board, with one as a backup. But North American—having their own troubles meeting weight requirements—quickly balked at the size and space requirements of two computers, and NASA agreed.

Another idea for increased reliability included having spare circuit boards and other modules on board the spacecraft so the astronauts could do "in-flight maintenance," replacing defective parts while in space. But the idea of an astronaut pulling open a compartment or floorboard, hunting for a defective module and inserting a spare circuit board while on

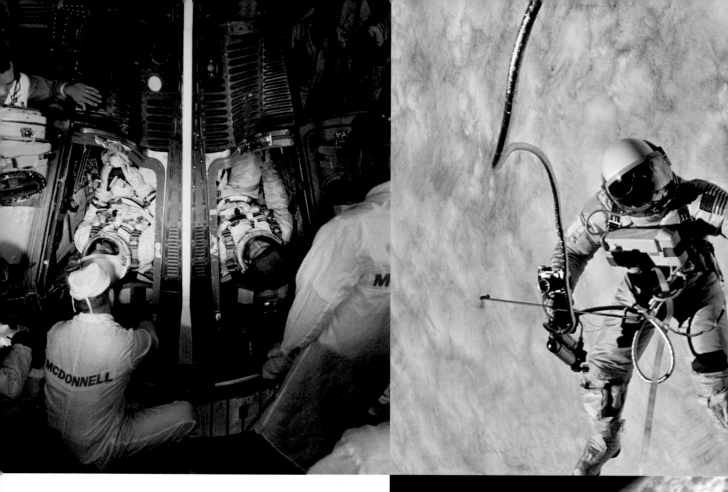

GEMINI 4

Gemini 4 brought James McDivitt and Edward White on a 4-day, 62-orbit, 98-hour flight which went from from June 3 to June 7, 1965. The mission included the first US spacewalk, by White, a critical task that would have to be mastered before landing on the Moon. Other objectives were to evaluate the procedures, schedules and flight planning for an extended flight, as well as try to conduct station-keeping and rendezvous maneuvers.

At the end of the 20-minute spacewalk, White was exuberant. "This is the greatest experience," he said. "It's just tremendous."

During the mission, McDivitt and White conducted 11 scientific experiments. One investigation involved spacecraft navigation using a sextant to measure their position using the stars, as the Apollo missions would need to do.

Clockwise from top left: Astronauts Ed White and Jim McDivitt are shown in the white room as they enter the Gemini 4 spacecraft atop the Titan launch vehicle at Cape Kennedy, Florida. Credit: NASA.

Ed White floats in the zero-gravity of space during the third orbit of the mission. White's face is shaded by a gold-plated visor to protect him from unfiltered rays of the sun. In his right hand he carried a Hand-Held Self-Maneuvering Unit (HHSMU) that gives him control over his movements in space. He was secured to the spacecraft by a 25-foot (7.6-m) umbilical line and a 23-foot (7-m) tether line. Credit: NASA.

Another view of White's EVA. Credit: NASA.

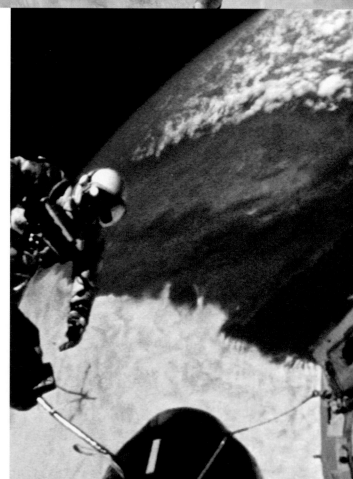

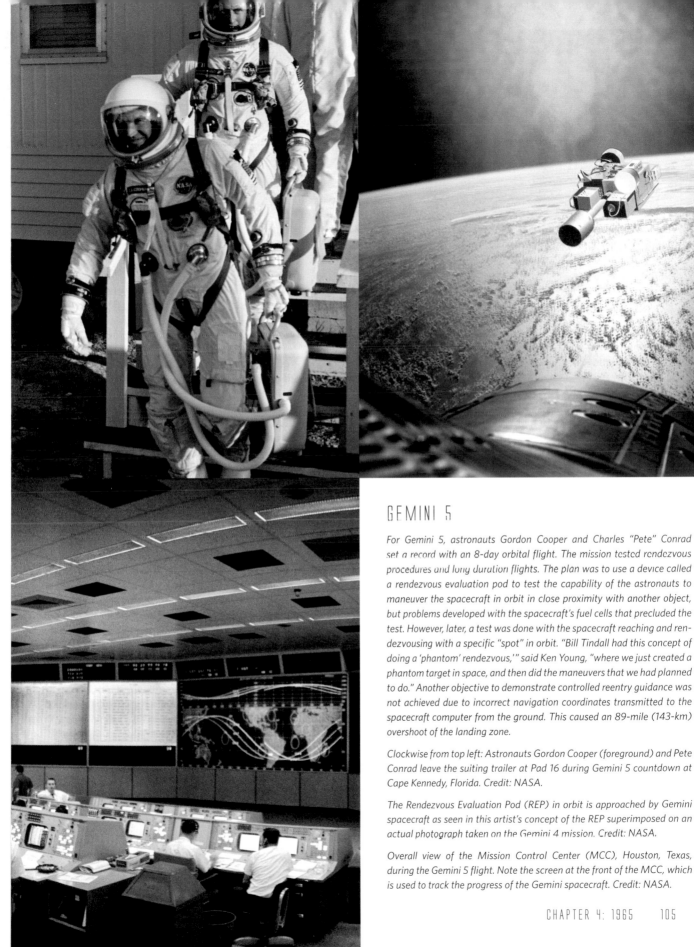

GEMINI 5

For Gemini 5, astronauts Gordon Cooper and Charles "Pete" Conrad set a record with an 8-day orbital flight. The mission tested rendezvous procedures and long duration flights. The plan was to use a device called a rendezvous evaluation pod to test the capability of the astronauts to maneuver the spacecraft in orbit in close proximity with another object, but problems developed with the spacecraft's fuel cells that precluded the test. However, later, a test was done with the spacecraft reaching and rendezvousing with a specific "spot" in orbit. "Bill Tindall had this concept of doing a 'phantom' rendezvous,'" said Ken Young, "where we just created a phantom target in space, and then did the maneuvers that we had planned to do." Another objective to demonstrate controlled reentry guidance was not achieved due to incorrect navigation coordinates transmitted to the spacecraft computer from the ground. This caused an 89-mile (143-km) overshoot of the landing zone.

Clockwise from top left: Astronauts Gordon Cooper (foreground) and Pete Conrad leave the suiting trailer at Pad 16 during Gemini 5 countdown at Cape Kennedy, Florida. Credit: NASA.

The Rendezvous Evaluation Pod (REP) in orbit is approached by Gemini spacecraft as seen in this artist's concept of the REP superimposed on an actual photograph taken on the Gemini 4 mission. Credit: NASA.

Overall view of the Mission Control Center (MCC), Houston, Texas, during the Gemini 5 flight. Note the screen at the front of the MCC, which is used to track the progress of the Gemini spacecraft. Credit: NASA.

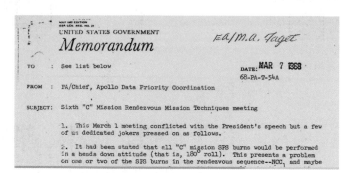

A portion of a "Tindallgram" from March 7, 1968, a memorandum from NASA's Bill Tindall.

Tindall went through the Lab's work with a fine-tooth comb and quickly grasped the key issues and clearly characterized the proposed fixes.

"Tindall came down hard on the Lab," said Ken Goodwin, who would join the MIT Instrumentation Lab in 1967. "He would come to MIT in what we called Black Friday Meetings. He would say, 'Don't do this, take this out. Why do you need that?'"

Tindall's oft-used motto was "Better is the enemy of good," and he'd tell the people at MIT, "Look, we just want the basic thing to work, no fancy displays, bells and whistles."

Of course, that rubbed some at the Instrumentation Lab the wrong way. "They were these gifted people, doing things that had never been done before," said Goodwin, "but here was Tindall saying, 'Hey, if we are going to meet schedule, we need to cut out the riffraff and get down to business.'"

Tindall documented all the issues and meetings in his notorious "Tindallgrams."

"Bill's Tindallgrams were written just the way he talks," said Osgood, "but he left out the swear words. For Bill, a swear word was an adjective. It just flowed with the rest of his conversation."

Most of Tindall's memos were long and detailed but also included well-placed constructive criticism. In his first MIT Tindallgram, Tindall defined the work at the Instrumentation Lab as "sloshing through the mud." Another early memo said, "Well, I just got back from MIT with my weekly quota of new ulcers, which I thought might interest you." Next came, "This is another of my gripping reports on the status of the Apollo spacecraft computer programs."

"Even though he kicked our butt, we ended up being in awe of what Tindall was doing," said Goodwin. "He found duplication in the software that made it extremely slow, he stopped the tendency for writing lengthy test programs to try out different things. He told us, 'This is not a research project; you have to get serious.'"

But as the Data Priority meetings progressed, Tindall's criticism of the Instrumentation Lab's effort softened as he came to a better understanding of the complexity of the tasks at hand. He did everything he could to provide additional manpower and equipment from MSC and other contractors, such as assigning NASA personnel to short-term residence at the Lab or arranging for the Lab to use some of the test facilities at the contractor sites.

Tindall helped institute certain restrictions, such as the requirement that the software be designed to use only 85 percent of the computer's total capability, always leaving 15 percent in reserve in case of an emergency or problem.

"The 85 percent was a software design goal, a performance target on the computer's hardware resources to execute the software," said Goodwin. "Due to the interrupt-driven executive design of the software, the computer's processing load was somewhat nondeterministic depending on what the computer was faced with during various points in the mission."

Another necessity of Tindall's was the refining of the breakout points for how the computer interrupted tasks. Tindall also astutely realized the complexities of the computer interfaces between all the systems on board the spacecraft. He oversaw what he called an interface control document, which detailed how one system interfaced with another and listed any change to one system and how it might affect another. Since everything with Apollo continued to happen at breakneck speed, it was hard to keep everyone in the loop. But if engineers or astronauts asked for a specific change in software or procedures, the interface control document could record and share that change among all systems managers.

Working toward landing on the Moon was "an incredible and audacious task," wrote David Hoag, the program manager for the Instrumentation Lab, "and the guidance equipment for the mission was created out of primitive principles, prolific imagination and a lot of hard work."

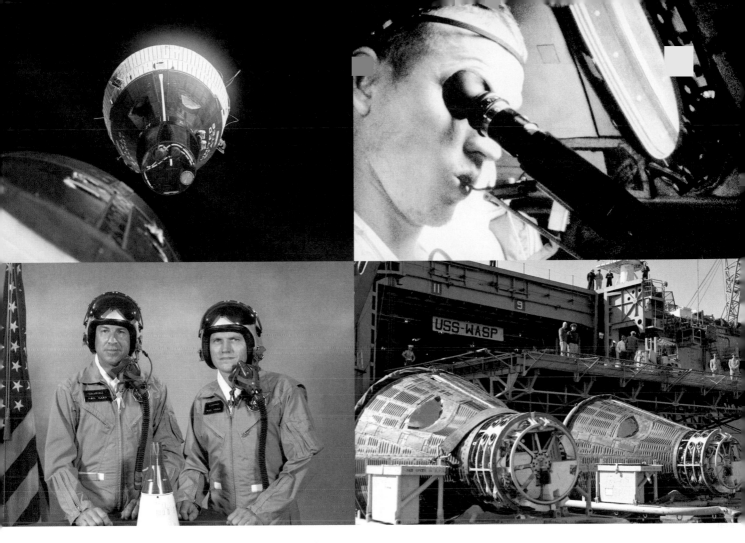

GEMINI 7 AND 6

This wasn't the original plan, but Gemini 7 ended up launching before Gemini 6, and the two missions combined for the first true rendezvous in space. Gemini 6's planned launch in October 1965 was thwarted when an unpiloted Agena spacecraft—one that Gemini 6 was supposed to conduct rendezvous maneuvers with—blew up during liftoff. NASA and the Gemini spacecraft contractor McDonnell Aircraft hatched the idea to launch Gemini 7 first, and to change Gemini 6's mission to rendezvous with Gemini 7 instead, which was scheduled to conduct the longest flight of a US spacecraft to date, fourteen days.

Gemini 7 launched on December 15, 1965, and astronauts Frank Borman and Jim Lovell demonstrated their ability to do experiments in a "shirt sleeve" environment and the lightweight pressure suits, and the capability of staying in space for a two-week period of time.

Eight days later, on December 12, Wally Schirra and Tom Stafford sat in their spacecraft waiting for launch when the engines started but cut one second after engine ignition because an electrical umbilical separated prematurely. This was the first time an astronaut mission was aborted after ignition start. The mission launched successfully three days later on December 15. Gemini 6 caught up to Gemini 7 and the two Gemini spacecraft flew in zero relative motion with between 45 and 120 feet (14 and 37 m) between them. Station-keeping maneuvers involved the spacecraft circling each other and approaching and backing off, where all four astronauts took turns flying in formation for over five hours. Gemini 6 returned to Earth on December 16, while Gemini 7 remained in Earth orbit and reentered two days later.

Top left: A photograph of the Gemini 7 spacecraft—nose toward camera—was taken from the Gemini 6 spacecraft during rendezvous and station-keeping maneuvers at an altitude of 163 nautical miles during orbit number 5, on December 15, 1965. Credit: NASA.

Bottom left: Astronauts Frank Borman (right), command pilot, and Jim Lovell for the Gemini 7 mission. Credit: NASA.

Top right: Frank Borman using the visual acuity device and a portable mouth thermometer during his experiment in space. Credit: NASA.

Bottom right: Gemini 7 (left) and Gemini 6 spacecraft meet up once again, this time at Mayport Naval Station near Jacksonville (Florida) after unloading December 20, 1965, from the carrier USS Wasp. Credit: NASA.

GEMINI 7 AND 6

Top left: A water-level view of Navy divers assisting Gemini 6 crew members Stafford and Schirra to open hatches after landing in the Atlantic. Credit: NASA.

Top right: The Gemini 7 spacecraft—side view—was taken from the Gemini 6 spacecraft during rendezvous and station-keeping maneuvers at an altitude of 157 nautical miles during orbit number 7, on December 15, 1965. The two spacecraft are approximately 45 feet (14 m) apart. Credit: NASA.

Bottom: Gemini 6 astronauts Thomas Stafford (left), pilot, and Walter Schirra, command pilot, are shown during suiting up exercises at Cape Kennedy, Florida. Credit: NASA.

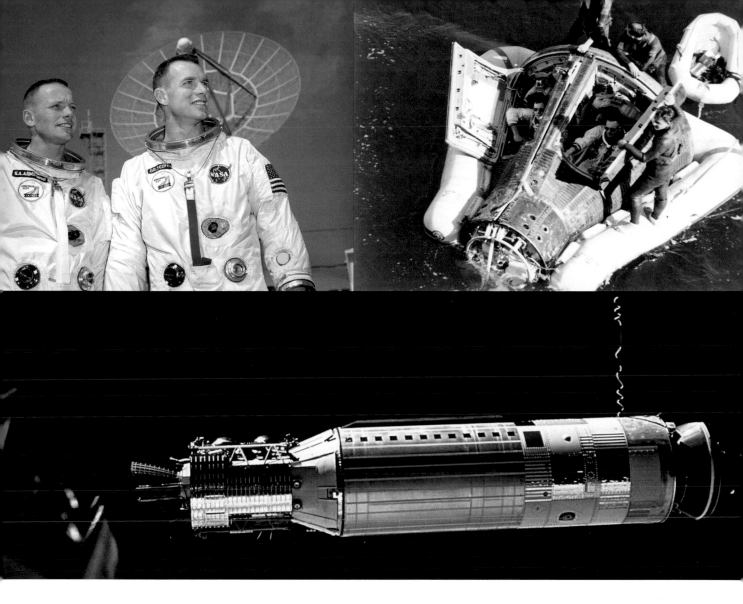

GEMINI 8

Gemini 8 launched on March 16, 1966, and astronauts Neil Armstrong and David Scott performed the first orbital docking of their spacecraft to an Agena target vehicle, the first mission to link two spacecraft together in Earth orbit. This milestone would prove vital to the success of future Apollo Moon landing missions, but problems ensued. During the rendezvous and docking maneuvers, the crew performed 9 maneuvers to rendezvous with the Agena, and docked during their fifth orbit. About 27 minutes after docking, the combined vehicles began spinning in a violent tumble and Armstrong disengaged the Gemini capsule from the Agena, thinking this action would solve the problem. But the Gemini spacecraft spun even more rapidly than when it was connected to the Agena, possibly exceeding a rate of one revolution per second. Armstrong and Scott managed to deactivate the Orbit Attitude and Maneuver System (OAMS) roll thrusters, as it had short-circuited, causing one thruster to fire continuously, causing the tumbling. To counteract the violent tumbling, the crew utilized all 16 reentry control system (RCS) thrusters to damp out the spinning, which succeeded in stabilizing the spacecraft.

Due to the premature use of the reentry control system, an immediate landing was required by Gemini safety rules, so the planned EVA and other activities were cancelled and the crew came back to Earth just over 10 hours after launch.

Top left: Astronauts Neil Armstrong (left), command pilot, and David Scott, pilot, during a photo session outside the Kennedy Space Center (KSC) Mission Control Center. Credit: NASA.

Top right: Astronauts Neil A. Armstrong and David R. Scott look calm and cool after their wild ride in space, sitting with their spacecraft hatches open while awaiting the arrival of the recovery ship, the USS Leonard F. Mason. The yellow flotation collar helps stabilize the spacecraft in choppy seas. Credit: NASA.

Bottom: A close view of the Agena Target Docking vehicle seen from the Gemini 8 spacecraft during rendezvous in space. Credit: NASA.

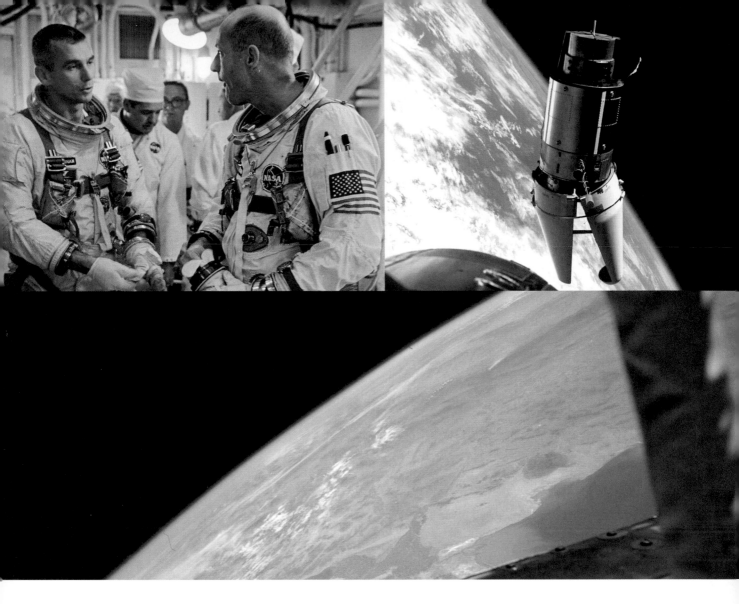

GEMINI 9

Gemini 9 was originally scheduled to launch on May 17, 1966, but was postponed when the Agena target vehicle failed to achieve orbit due to a booster failure. The replacement Augmented Target Docking Adapter (ATDA) was launched successfully into Earth orbit on June 1, but telemetry indicated that a shroud covering the adaptor had failed to jettison properly. Gemini 9 was supposed to launch that same day, but ground equipment failure resulted in a postponement until June 3.

Astronauts Tom Stafford and Eugene Cernan were able to come within 26 feet (8 m) of the ATDA on the third orbit, and confirmed the launch shroud on the ATDA had failed to deploy and was blocking the docking port. The flight plan was then revised to include two passive rendezvous maneuvers instead of the docking.

Two days later, Cernan conducted a 2-hour spacewalk, the longest to date. However, he found the tasks in zero G took "four to five times more work than anticipated," overwhelming Cernan's environmental control system and causing his faceplate to fog up, limiting his visibility. Radio transmissions were also garbled between Cernan and Stafford, cutting the spacewalk short.

Top left: Astronauts Eugene Cernan (left), pilot, and Thomas Stafford, command pilot, discuss the postponed Gemini 9 mission just after egressing their spacecraft in the white room atop Pad 19. Credit: NASA.

Top right: The Augmented Target Docking Adapter (ATDA) as seen from the Gemini 9 spacecraft during one of their three rendezvous in space. Failure of the docking adapter protective cover to fully separate on the ATDA prevented the docking of the two spacecraft. The ATDA was described by the Gemini 9 crew as an "angry alligator." Credit: NASA.

Bottom: Gene Cernan took this picture from the Gemini 9 spacecraft, over California, Arizona and Sonora, Mexico, during his EVA. Credit: NASA.

GEMINI 10

Launched on July 18, 1966, the Gemini 10 mission was the first flight to conduct two rendezvous and docking tests with the Agena target vehicle. Astronauts John Young and Michael Collins were on board, and Collins became the first person to visit another spacecraft in orbit. The crew performed two EVAs, along with fifteen scientific, technological and medical experiments. The mission also set a new altitude record for human spaceflight, reaching 475 miles (764 km). Young and Collins returned to Earth on July 21, 1966.

Clockwise from top left: The Gemini 10 spacecraft launched on July 18, 1966. A time exposure creates the illusion of multiple rocker arms. Credit: NASA.

Agena Target Docking Vehicle docked to Gemini 10 spacecraft. The "glow" is from the Agena's primary propulsion system. Credit: NASA.

Twelve-year-old Billy Doyle of Virginia Beach, VA, shakes hands with Mike Collins aboard the recovery ship USS Guadalcanal. At right is John Young. Billy represented 41 youngsters permitted aboard the Guadalcanal to witness the recovery with their Naval fathers or close relatives, marking the first time dependents were permitted aboard a ship during a Gemini recovery operation. Credit: NASA.

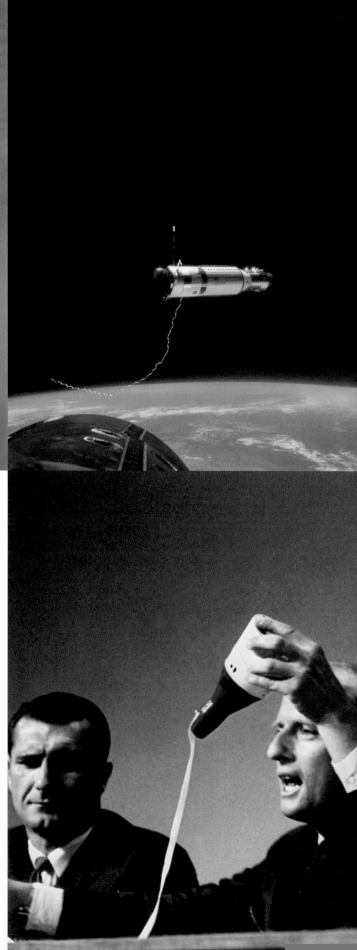

GEMINI 11

Gemini 11 launched on September 12, 1966, carrying astronauts Pete Conrad and Richard Gordon. The 3-day mission achieved a first orbit rendezvous and docking with the Agena target vehicle, which had been launched an hour and a half before Gemini 11. Each astronaut then conducted two docking exercises with the Gemini-Agena Target Vehicle (GATV). Gordon conducted a scheduled 107-minute EVA, but found the tasks to be exhausting and heavy perspiration inside his spacesuit helmet obscured his vision and finally blinded his right eye. Conrad ordered him to cancel his work and return to the cabin.

On September 14 the crew fired thrusters to raise the docked spacecraft to over 850 miles (1,368 km) above Earth, a record altitude for an astronaut mission that would stand until Apollo 8 went to the Moon.

Clockwise from top left: The Agena Target Docking Vehicle launched from Launch Complex 14 at Cape Canaveral on September 12, 1966. Credit: NASA.

The Agena Target Docking Vehicle is tethered to the Gemini 11 spacecraft during its 31st revolution of Earth. Credit: NASA.

Pete Conrad (left), command pilot, and Dick Gordon (right), pilot, demonstrate tether procedure between their Gemini 11 spacecraft and the Agena Target Docking Vehicle at the post flight press conference. They use models of their spacecraft and its Agena to illustrate maneuvers. Credit: NASA.

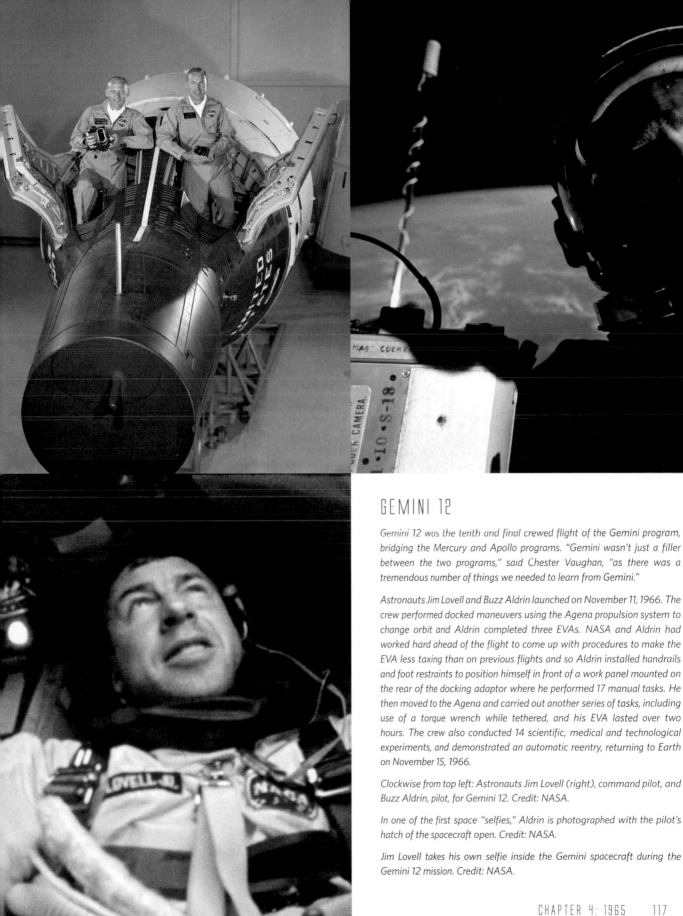

GEMINI 12

Gemini 12 was the tenth and final crewed flight of the Gemini program, bridging the Mercury and Apollo programs. "Gemini wasn't just a filler between the two programs," said Chester Vaughan, "as there was a tremendous number of things we needed to learn from Gemini."

Astronauts Jim Lovell and Buzz Aldrin launched on November 11, 1966. The crew performed docked maneuvers using the Agena propulsion system to change orbit and Aldrin completed three EVAs. NASA and Aldrin had worked hard ahead of the flight to come up with procedures to make the EVA less taxing than on previous flights and so Aldrin installed handrails and foot restraints to position himself in front of a work panel mounted on the rear of the docking adaptor where he performed 17 manual tasks. He then moved to the Agena and carried out another series of tasks, including use of a torque wrench while tethered, and his EVA lasted over two hours. The crew also conducted 14 scientific, medical and technological experiments, and demonstrated an automatic reentry, returning to Earth on November 15, 1966.

Clockwise from top left: Astronauts Jim Lovell (right), command pilot, and Buzz Aldrin, pilot, for Gemini 12. Credit: NASA.

In one of the first space "selfies," Aldrin is photographed with the pilot's hatch of the spacecraft open. Credit: NASA.

Jim Lovell takes his own selfie inside the Gemini spacecraft during the Gemini 12 mission. Credit: NASA.

CHAPTER 5

1966

Man. Moon. Before 1970.

DAVE SCOTT, Apollo astronaut

NASA, IT SEEMED, WAS ON A ROLL.

As 1966 began, half of the Gemini flights had been completed, and the program was well on the way to accomplishing all its major goals by the end of the year. Several Apollo test flights were scheduled for the year and, if all went well, there were hints that NASA could be on course for the first crewed Apollo missions by the end of the year.

The first Apollo test flight, called Apollo Saturn-201, was scheduled for February. Dottie Lee took particular interest in this flight, because it would be the first true test of her Apollo heat shield. She knew that every Apollo mission would return in the same way it left Earth: in a searing mass of heat and flame. With the return velocities from the Moon predicted to be 25,000 miles (40,000 km) per hour, the Command Module (CM) plunging through the atmosphere would experience greater heat and stresses than any previous spacecraft, reaching temperatures of more 3,000°F (1,649°C). Just inches would separate the crew from the fiery exterior. A new type of protective heat shield was required.

Command Module 009, after its flight in February 1966, at the North American Aviation facility in Downey, California, for postflight testing. Image courtesy of Rich Manley.

Left: AS-201, the first Saturn IB launch vehicle lifts off from Cape Canaveral, Florida, February 26, 1966. Credit: NASA.

Charles A. Bassett II (left) and Elliot M. See Jr. were supposed to fly the Gemini 9 mission. This image was taken in January 1966. Credit: NASA.

Dr. Robert R. Gilruth (far right) introduces the Apollo 1 crew during a press conference in Houston. From the left are astronauts Roger Chaffee, Edward H. White II and Virgil I. (Gus) Grissom. Credit: NASA.

Lee and her colleagues conducted a series of calculations to measure the thermodynamic characteristics of reentry and—with wind tunnel tests, data from the Mercury missions and mathematical predictions—Lee and her colleagues computed the performance required for a heat shield for Apollo.

"From our tests, we knew the only material that could get us back from the Moon was an ablater," she said. They had tested several different types of protective materials (such as Teflon) but finally determined the best option was an ablative material that would melt and erode away the accumulated heat of atmospheric reentry while protecting the spacecraft. And the larger Apollo CM needed a new type of ablative material. Lee worked with several companies, and after three years of research, NASA chose Avco Corporation in Lowell, Massachusetts. They had developed a new type of epoxy resin called Avocoat, with special fiberglass additives. The resin would be injected into a steel honeycomb mesh and bonded to the shell of the CM. Lee traveled to the Avco plant and watched as technicians used special heated injection guns to meticulously fill the four hundred thousand tiny holes in the honeycomb matrix with the ablative material. Lee touched the ablator right out of the gun; it felt like a warm putty entwined with fibers. After it hardened, this would be the material that would keep the interior of the spacecraft comfortable and safe while an inferno raged outside.

Lee anticipated the February 26 flight because so far, all the calculations and tests had been theoretical. Nothing was better than real, hard data. But the launch was a huge disappointment. Several malfunctions occurred, mostly minor, but three were serious. First, the propulsion system on the Service Module (SM) malfunctioned. Second, an electrical fault caused a loss of steering control. And third—the problem that affected Lee most—a short-circuit caused the loss of the data on the heat shield. She knew she'd have to wait several months for the next launch to get the data she so desperately wanted.

Back in her office the following Monday morning, February 28, Lee was still feeling disappointed when she received a phone call from her husband, John, with shocking news. Two astronauts, Elliot See and Charlie Bassett, had been killed earlier that morning in a plane crash. They were flying together in a T-38 jet amid deteriorating weather conditions of fog and snow, heading to St. Louis, Missouri. Upon making their landing approach at Lambert Field in St. Louis, See, an experienced naval aviator, misjudged the approach, coming in too low and too fast. They were killed instantly when they crashed into the nearby McDonnell Douglas factory, where the Gemini spacecraft were being manufactured. See and Bassett were supposed to fly the upcoming Gemini 9 flight and were on their way to McDonell Douglas for training. Part of the structure was severely damaged from the crash, and had the plane hit the other side of the factory, it could have slammed into the assembly line, perhaps killing hundreds of McDonnell's workers and destroying at least two Gemini spacecraft.

The accident sent shockwaves of grief through NASA. Dottie Lee had always known that getting to space was risky. She'd seen her share of rocket failures and experienced other setbacks. And she also knew going to the Moon was especially dangerous—in the back of her mind, she had considered the high likelihood of losing astronauts before reaching that goal. But losing them here on Earth seemed especially tragic. Not getting her heat shield data now seemed trivial.

But Lee had no doubt NASA would continue toward the Moon, because everyone she worked with was motivated to succeed and to do it right.

Lee would have to wait until August until the next test, when a boilerplate version of the CM was launched on a Saturn IB, with trajectories that enabled a simulation of the lunar reentry speeds. When they got the data, it showed the outer surface of the CM reached 2,700°F (1,500°C). The interior remained a comfortable 70°F (21°C).

Another test flight in July verified the S-IVB stage design for Saturn V was restartable, and that meant plans could proceed for the first crewed flight, Apollo Saturn-204, or AS-204. NASA announced astronauts Gus Grissom, Ed White and Roger Chaffee as the first crew for Apollo. According to NASA's announcement, the plans for the astronauts' test flight in Earth orbit included the intent "to verify spacecraft crew operations and CSM subsystems performance for a mission of up to 14 days." These astronauts were three of NASA's best and brightest: Grissom, the second American in space and a veteran astronaut; White, a West Point graduate who conducted the first American spacewalk on Gemini 4; and Chaffee, a rookie but considered one of the agency's most talented engineers. The astronauts and ground crew continued training and testing in order to meet the optimistic goal—but in the meantime, problems, glitches and disagreements on the ground kept everyone at NASA occupied.

WHEN ELBERT KING CAME TO WORK AT
the Manned Spacecraft Center (MSC), he just assumed everyone knew the reason that NASA was going to the Moon. To King, the justification for Apollo was as clear as when he looked through his geologist's hand lens to see close-up details of sandstone or shale: NASA was landing on the Moon to collect rocks because that's where all the scientific information could be found. But King quickly discovered lunar samples were not the dominant aspect of Apollo. In fact, science held a low standing on the Apollo priority list.

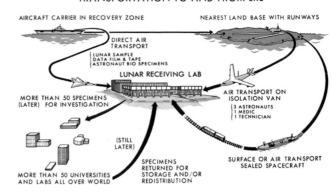

TRANSPORTATION TO AND FROM LRL

Plans for the transportation of lunar materials to and from the Lunar Receiving Laboratory. Credit: NASA.

Nonetheless, initial plans were unfolding for the astronauts to collect a few Moon rocks. King and his new colleague, geochemist Don Flory, had been hired to design airtight sample return containers for the lunar materials. But the two scientists realized no one was giving any forethought to how the rocks should be handled once they were on Earth.

King and Flory showed up in James McLane Jr.'s MSC Facilities Office, expressing their concerns.

"Does anyone around the Center have a small vacuum chamber where we can open these boxes?" they asked. "The scientific integrity of the samples would be compromised if they are exposed to Earth's atmosphere. And what are the plans for how are they going to be studied and then stored?"

That conversation birthed the concept of MSC's Lunar Receiving Laboratory (LRL). But what began as a seemingly straightforward idea of building a facility to store and study rocks from the Moon quickly became a power struggle between engineers who would build the facility, scientists who wanted to study the rocks and members of the medical community who felt they needed to save the world from biological disaster—not to mention even more squabbling between the various governmental agencies and politicians.

McLane and King ended up in the middle of it all. To achieve the goal of completing the laboratory in time to receive samples from the first Moon landing, MSC director Robert Gilruth instituted a board of experts in May 1966 that was given the "authority to make policy decisions in minimum time." Some of the committee members were appointed by NASA Headquarters and were primarily scientists who had been selected as principal investigators for the proposed experiments and study of the lunar samples.

Astronaut Michael Collins on the right during a tour of Lunar Receiving Lab (LRL) at MSC with LRL Administrative Assistant Richard Wright on the left. They are among the secure glove boxes that would be used to safely study lunar materials. Credit: NASA.

The initial plan called for a small clean room of approximately 100 square feet (9 sq. m) where the sample boxes could be opened under vacuum conditions and repackaged for distribution to various researchers. But some scientists and NASA officials concluded just a single room wouldn't be sufficient and quickly came up with a plan for a 2,500-square-foot (232-sq.-m) research facility where the lunar samples would not only be stored but studied as well. After more discussion, an 8,000-square-foot (743-sq.-m) version was proposed.

Working with the scientific advisory committee to develop a workable plan for the ever-growing and changing proposed facility turned out to be an interesting challenge for McLane and his team.

"The biggest challenges were political," McLane said. "All the scientists involved in studying the samples had laboratories of their own. They didn't want to do anything unless it was going to benefit their facility back home. Others were suspicious that we were trying to appropriate activities that weren't in the Manned Spacecraft Center's charter at the expense of other NASA centers."

McLane found it challenging to get everyone involved to cooperate and agree on the initial receiving procedures. A few of the proposed experiments, such as those to determine the radiation properties of the lunar samples, were very time-dependent. Therefore, it became evident that the facility and equipment required to perform those experiments would have to be located near the point where the samples were first available. That point was Houston, and McLane said it particularly rankled some of the scientists to see new state-of-the-art facilities and equipment being located at Houston rather than at their home laboratories.

"I had never worked with high-level scientists before, and our advisory committee usually consisted of people who were at the level of principal assistants to Nobel Prize winners," McLane said. "Overall, it was a great group to work with, with one important exception. They each reserved the right to change their mind."

Frequently, a previously settled contentious issue was brought up again some weeks later and a different solution was proposed, with the instigator pleading, "Well, I was just wrong before," or "I changed my mind." McLane felt the committee was just ignoring an extremely tight schedule, as well as reality.

For example, one issue was whether to use glove boxes or a closed container with mechanical manipulators to work with the Moon rocks. It took many discussions and debates to decide, and the ultimate decision would make a big difference in what direction the engineers needed to go for building the lab. With a limited time to decide, the glove box idea won out.

But in the mid-1960s, lunar scientists still debated whether the Moon's craters were volcanic or created by impacts. No one knew the elements from which the Moon was composed. Not knowing the answers to almost any scientific question about the Moon meant that planners for the Lunar Receiving Lab had to be wide-ranging in their preparations. King recognized the quality of research on the initial samples needed to support the increased scientific opportunities being proposed for succeeding Apollo flights.

McLane was also surprised at all the different scientific speculations that took place. Astrophysicist Thomas Gold, a member of the Presidential Science Advisory Committee for Space, had long proposed the Moon was covered with several hundred feet of lunar dust and suggested a spacecraft landing on the surface would be swallowed up into the regolith. Other scientists proposed the Moon rocks—originating in hard vacuum and bombarded with radiation and meteorites—might catch fire or explode if exposed to Earth's rich atmosphere.

"The speculations by good, smart, reputable people were just unlimited," said McLane. "But I guess they were trying to think of all the possibilities. We were fortunate that no one forced us to plan for any of these extreme speculations. Overall, our advisers did a good job."

But then at one of the meetings in Washington, DC, to meet with advisers at NASA Headquarters, a scientist from the Public Health Service arrived and asked how NASA was going to protect against contamination of the Earth by lunar microorganisms.

McLane said the initial reaction by everyone else was, "What?"

The Lunar Receiving Laboratory after construction was completed in 1967. Credit: NASA.

Since the 1950s, a small group of scientists had discussed the remote possibility that any lunar samples brought back to Earth might contain deadly organisms that could destroy life on Earth. Even the spacecraft and the astronauts themselves could possibly bring back nonterrestrial microorganisms that could be harmful. Several governmental agencies, including the Department of Agriculture, the US Army and the National Institutes of Health got wind of this idea—perhaps blowing it a little out of proportion, McLane felt—and NASA was forced to take action to prevent a possible biological disaster.

"The 'lunar bugs,' as we called them," said McLane, "well, nobody that I worked with really believed there was life on the Moon, especially something that might affect people—make them sick or kill off our civilization, that sort of thing."

McLane said that the first time head astronaut Deke Slayton heard about this, he just about "flew out the window."

"No way is somebody going to step in and put these restraints on the program," Slayton fumed. "It's difficult enough to just fly to the Moon without all these precautions about contamination."

Dr. Charles Berry was the director of medical research and operations at MSC and felt pressures from both sides of the contamination debate. On the one hand, he knew some concern was warranted, but as a representative of everything going on in Houston, he "had been charged by NASA to say that we were indeed not going to bring back lunar plague."

Typical astronaut living quarters in the Crew Reception Area of the Lunar Receiving Laboratory, Building 37, at the Manned Spacecraft Center in Houston, Texas. Credit: NASA.

So, while NASA believed the chance of lunar back-contamination was negligible, they were forced to institute a quarantine program because of the medical community's and US agencies' concerns. King and Flory's team determined the initial handling and examination of the lunar samples would have to be performed in sophisticated, hard-vacuum chambers, upping the cost even more for the LRL. McLane and fellow MSC officials expressed concern on how the scope and the price tag of the lab continued to rise. But eventually, the scientific justification and worries of public fears won out over budget.

"We had meetings with the surgeon general of the US," McLane said, "and he took the attitude, 'How much is the Apollo program going to cost—$20 billion or so? I don't think it is outlandish to set aside one percent of that to guard against a great catastrophe on Earth.' We said that we would take on the challenge of guarding against organisms, but the surgeon general would have to justify the increased costs to Congress. And he did. That settled that, so we developed a scheme and it was approved. Everyone had to accept it; there wasn't any choice."

That changed the entire complexion of what McLane and his team had to accomplish before astronauts could go to the Moon. The initial small clean room would now have to be a research lab plus a quarantine facility, with living accommodations for the astronauts and several other people during a three-week period after each lunar mission. Plans for the facility grew to an 86,000-square-foot (8,000-sq.-m) structure that would cost more than $9 million. From the beginning, King and Flory were highly involved with designing the scientific experiments and facilities but now had to take on developing the detailed, initial tests on the rocks *and* the astronauts that had to be done quickly behind absolute biological barriers to test for any contamination.

Gilruth assigned his assistant, Dick Johnston, as operations manager for the LRL and hired Dr. Persa R. Bell, a nuclear physicist, as chief of MSC's Lunar and Earth Sciences Division to make sure the new lab met all the mounting requirements. Although many engineers at MSC felt the elaborate precautions for contamination provided superfluous impediments to the Apollo program, the scientific uncertainty meant NASA could not afford to take any risks. If they were wrong about the reality of "lunar bugs," the consequences could be disastrous.

ALTHOUGH SCIENCE PLAYED A ROLE IN almost every aspect of Apollo, NASA needed as much science as they could get in regard to figuring out where and how to land on the Moon. To truly determine the conditions the Lunar Module (LM) and the astronauts would face upon landing, precursor reconnaissance observations of the Moon were necessary. With the expertise of the Jet Propulsion Laboratory (JPL) and Langley Research Center, NASA began a broad program of lunar landers and orbiters to prepare for Apollo. During the 1960s, NASA launched a fleet of twenty-two robotic spacecraft, blazing a trail toward the Moon. While the early missions had not been designed initially to support Apollo, successive spacecraft were retooled to gather more data for mapping the lunar surface. The Ranger missions crash-landed (purposely), while the Surveyors soft-landed and the Lunar Orbiters circled the Moon. The missions sent back significant data and pictures of the Moon, providing more detail than astronomical observations could at the time.

The three successful Ranger missions (out of nine total) took images with enough detail to show that a lunar landing was likely quite feasible, but the sites would have to be carefully chosen to avoid craters and big boulders. In February 1966, the Soviet Union soft-landed the *Luna 9*. The US soon followed with six Surveyor spacecraft, proving the lunar surface could easily support the impact and the weight of a small lander. The Surveyors included television cameras that beamed back grainy but eye-level views of the Moon.

The Lunar Orbiters' five missions showed that objects could successfully enter lunar orbit and provided the best images yet of the lunar surface. The onboard cameras could capture surface details as small as 4 feet (1.25 m) across. A technological marvel for 1966, the onboard camera took images using 70-millimeter film; it was developed automatically using a process similar to that of the Polaroid Instant Film camera, and an electron beam would then scan each developed image before transmitting the photos back to Earth. It used analog radio signals, similar to how television satellites in the mid-1960s sent signals to TV stations.

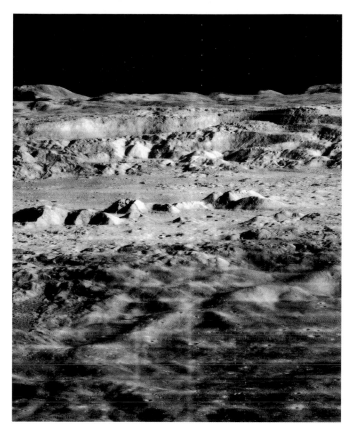

A composite image from the Lunar Orbiter II mission in 1966, showing a close-up view of the floor of the Moon's crater Copernicus. Credit: NASA.

Through all five missions, twenty potential Apollo landing sites were targeted, with 99 percent of the lunar surface mapped at low resolution. Additionally, sensors on board the spacecraft indicated that radiation levels near the Moon would pose no danger to the astronauts. But analysis of the spacecraft orbits found evidence of what scientists called "mascons"—mass concentrations or lumps of high-density regions below the surface of the Moon that perturbed the motion of spacecraft in lunar orbit. NASA engineers would need to take these orbital perturbations into account, and more study was needed of how they might affect the Apollo missions.

"We had to verify the calibration of the transducers and make sure all the data from the instrumentation on board the spacecraft would come through the telemetry to the ground," said Manley. "The instruments tell you how—and if—your system is working. If nothing else, you need a gas gauge to tell you how much propellant your thrusters are using."

Manley and his fellow test engineers performed testing and troubleshooting on the instrumentation for all eleven systems, including propulsion, guidance and control, communications and environmental control. The test engineers at North American needed to know the details of them all.

"The company did all the specialized training we needed on-site," Manley said, "and this was the best education I ever got, bar none. We had to know how all the systems worked so we could make sure all the calibrations were correct."

In total, approximately fifteen hundred measurements were transmitted from the spacecraft, and many were measured at least once a second (such as temperature and pressure sensors on every tank and the numerous thermocouples buried within the ablative heat shield). Other systems, like the biomedical sensors for astronaut health information (such as heart rate), were measured two hundred times a second.

"Our whole objective was to test the equipment to make sure it was capable of doing what it was designed to do," said Manley, "and we had to make sure all the data streams came through the telemetry. They couldn't fly with an instrumentation cable trailing from the spacecraft down to Earth, so it all had to be telemetered down."

The concept of test engineering was a relatively new profession, so they sometimes had to invent the technology and equipment to do their testing. "Nobody had any experience in this because it was so groundbreaking," said Manley. "From my standpoint, the opportunity to work on Apollo came because I just happened to have the right background at the right place at the right time."

While working on Apollo was undoubtedly exciting and challenging, the pressures of meeting schedule and performance demands from NASA meant office politics often came into play.

Test facilities at North American Aviaition. Image courtesy of Rich Manley.

"The slightest little things became a mountain from a molehill," Manley said. "I didn't like the politics, because I'm a doer and I don't like to play games. I'd get in trouble sometimes because if I ran a test without the holy-water blessings of the management, I'd get my backend chewed."

Getting the first manned Apollo spacecraft—known as *Spacecraft 012*—ready for the first flight, Manley answered questions for North American and NASA engineers, and he regularly saw Grissom, White and Chaffee and other early Apollo astronauts as they came to Downey to check out their ride to space.

"They followed the construction of their spacecraft," said Manley, "and if they wanted any changes made, they told North American and NASA design engineers. They weren't encumbered with a lot of paperwork. We'd provide real-time data for the astronauts or anyone who needed it, and our function was to give them a quick way of getting information."

North American had planned to ship *Spacecraft 012* from Downey to Kennedy Space Center in early August 1966, but continued problems with a glycol cooling system in the environmental control unit caused a delay. Since it was a Block I, which was a heavier version of the CSM, *Spacecraft 012* could only be flown in Earth orbit. The spacecraft arrived in Florida on August 26, 1966, with much fanfare. It was protected by a special cover, emblazoned with "Apollo One," which was the astronauts' preferred term for their mission, although NASA still officially called it AS-204.

Left: Workers at North American Aviation in Downey, California, build Spacecraft 009. Credit: NASA.

Service Module RCS Quad Panel and tanks. Credit: NASA.

Reaction Control System during a test at the Manned Spacecraft Center. Credit: NASA.

OTHER SYSTEMS WERE THREATENING

delays of the first crewed Apollo launch, and one issue in particular caused considerable headaches at MSC since Apollo couldn't fly if the problem persisted. During vacuum chamber tests, the small steering rockets for the Apollo spacecraft started blowing up.

"The reaction control engines are on the outside of the SM and the LM," said Norman Chaffee, "hanging just inches from the side of the spacecraft outside wall. So, if one blows up, it acts like a hand grenade, creating a lot of shrapnel."

A thruster explosion during a spaceflight could be catastrophic. Therefore, the explosive nature of this problem meant that everyone at NASA—in particular the astronauts—took a high interest in the details of the issue, especially when the Reaction Control System (RCS) thrusters kept exploding during repeated tests, for months on end. Management in the Propulsion and Power Division instigated a twenty-four-hour-a-day test regimen to figure it out.

On the SM, sixteen of these metallic, bell-shaped thrusters hung in clusters of four, spaced evenly around the spacecraft's outside barrel. These thrusters allowed for orientation of the vehicle in any desired direction just as an aircraft's elevators, ailerons and rudder control pitch, roll and yaw. Marquardt Company in Van Nuys, California, manufactured the thrusters and they, too, were working frantically to solve the problem of unexpected explosions.

After several months of dead ends in trying to understand the problem, Norman Chaffee and Henry Pohl wondered if they could find a way to look inside the combustion chamber as the thrusters fired to see what was going on. The chamber was slightly smaller than a soda can and made of high-melting point metals, but they constructed a new combustion chamber out of clear plexiglass. They set up the see-through thruster and a camera inside a vacuum chamber and fired the thruster while taking high-speed photos—twenty-five thousand frames a second—of the initial few milliseconds of firing. They gathered temperature and pressure data as well.

"The heat input from one of these short pulses was low enough that the plexiglass would last quite a long time," Chaffee said. "And only if we turned the engine on and just let it run would it eat up the plexiglass fairly quickly."

But it turned out they didn't see much, and even with higher-speed film, they didn't find any anomalies in the thruster. But the plexiglass thrusters blew up too. "We blew up lots and lots of plexiglass chambers without getting the data that we wanted," said Chaffee, "but we finally documented the formation and buildup of a gooey, yellowish-brown gunk that would collect in the rocket engine."

The gunk seemed to only form when the fourteen-millisecond thruster firings came in short spurts with long periods of time in between; occasional long, extended bursts appeared to eliminate the gunk. For reasons they could never adequately predict, at some point a critical mass of the gunk would collect and a short pulse would cause detonation.

"When it got to critical mass, it didn't just burn," said Chaffee, "it detonated. So it was an actual explosion."

To find out more, Pohl arranged to put the RCS thrusters in a long-duration test in a vacuum chamber in MSC's Thermo-chemical Test Area. They conducted a fourteen-day simulated mission, firing the thrusters in a manner that duplicated a spaceflight to the Moon and back, with about 175,000 firings of each thruster. But during the test, technicians noticed the inside of the vacuum chamber was getting coated with the same gunk that they'd seen inside the engines.

"We went in the vacuum chamber and took some samples of this stuff," Pohl said. "One of the guys put some gunk on an anvil and tapped it with a hammer. It popped like a cap pistol going off. We knew we had to get this stuff tested to find out what it was."

Pohl put some gunk in a vial and told Chaffee to take it over to the transportation office and ship it up to the Bureau of Mines in Pittsburgh for analysis. Chaffee soon returned to Pohl's office, not sure what to do.

"Can't ship it," Chaffee said. "We've got to find out whether it's a Class C explosive or not before we can ship it."

"Now how in the world are we going to find out if this is an explosive unless you get it to the Bureau of Mines for them to test it?" Pohl asked.

Chaffee knew this impossible circular discussion didn't have an answer.

"Well, I know how," Pohl said determinedly. "Put this vial in your briefcase and get on an airplane and get up there tonight."

Chaffee took the small sample, boarded a commercial flight and without incident made it to Pittsburgh. The Bureau of Mines determined the gunk was hydrazinium nitrate.

LM RCS subsystem in a vacuum chamber test in Building 353 at the Manned Spacecraft Center in Houston. Credit: NASA.

"You have to hit that stuff pretty hard to get it to detonate, but it detonates," Pohl said, "and that's what would happen in the combustion chambers. It would build up in there and explode. So we knew we had to figure out a way to eliminate that from happening."

In the meantime, the fourteen-day test was still under way and Pohl wanted to continue it to get as much data as he could. Just a few nights later, one of the technicians arrived on shift, worried.

"He came in about 11:00 at night and said that stuff must be getting out of the chamber because it was all over the cars outside in the parking lot," said Pohl.

The Apollo Simulators at Kennedy Space Center. Credit: NASA.

By late fall, the Command and Service Modules (CSM) simulator had lost its status as a lemon. The crews were training both in Houston and at the Cape, and Hughes knew that Grissom, White and Chaffee were going to be a well-trained crew for the first flight of Apollo. Whether they would get off the ground before the end of the year remained to be determined.

THE GEMINI PROGRAM CONCLUDED IN

November 1966, with the program successfully building a technological bridge between Mercury and Apollo. The missions simulated the length of a lunar round-trip, and astronauts learned how to move about and work during extravehicular activities (EVAs) in their spacesuits. The rendezvous concept was perfected, and the astronauts, flight controllers and mission planners became comfortable with these orbital maneuvers that would be necessary in lunar orbit.

One nagging issue with Gemini needed to be resolved, however: another thruster issue.

"We were experiencing a decay in the amount of thrust from the small ablative thrusters in Gemini," said Chester Vaughan. "They were supposed to put out 25 pounds (11 kg) of thrust, but over the course of the mission, it would gradually decay and get down to 2 to 3 pounds (1 to 1.4 kg) of thrust. That happened in at least eight out of the ten Gemini flights."

There was no direct measurement available from the spacecraft showing the amount of thrust at any specific time, but Jerry Bell and Ken Young from the rendezvous team could see evidence of the low thrust in looking at data from a strip chart recorder that tracked all the thruster firings. Gerry Griffin in Mission Control tracked the problem by analyzing the motion of the vehicle.

On the whole, the flight controllers were able to work around the problem because of a redundant system, but everyone wanted to figure out what was going wrong on Gemini

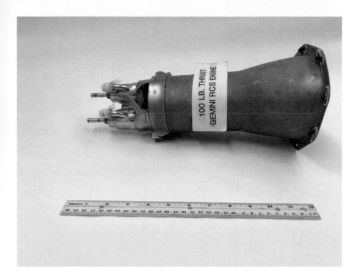

A 100-pound (45-kg) thrust Gemini reaction control engine, used for on-orbit positional maneuvering and control. This control system was called the Orbital Attitude Maneuvering System (OAMS) and was located in the aft structure of the orbital vehicle, which was separated prior to Earth reentry. Credit: NASA/Mike Salinas and Norman Chaffee.

The crew of Apollo 1, during recovery training. Credit: NASA.

because Apollo had the same type of maneuvering thrusters. While the Gemini flights were still ongoing, Vaughan had asked the program manager, Charles Matthews, if they could put instrumentation on some of the later Gemini flights to get more data. Matthews refused, saying it wasn't an issue. But immediately after the final Gemini mission, Matthews sent a memo to Vaughan saying he and his team needed to figure out the problem, and they had one month in which to do it.

Vaughan went to Matthews and said, "How come you didn't let us put that instrumentation on Gemini so we could figure this out?"

Matthews replied that frankly, if he had allowed the instrumentation, he'd have everyone wondering what the problem was and then they'd ask all sorts of questions about whether Gemini could still fly safely. Since it was more of a nagging problem, Vaughan and Matthews knew it wasn't a safety issue. But now Matthews provided Vaughan additional resources, sending him and his team to St. Louis to the McDonnell facility—where Gemini had been built—for an exhaustive set of tests. They got lucky. They were able to simulate the problem on the first test run, finding the small orifices in the thrusters were getting clogged by an excess of iron in the oxidizer.

"I think one of the reasons we were so successful in Gemini and then Apollo was that we worked hard," Vaughan said. "But we were lucky, too, and we'll take luck every time. But anytime we had a failure, we didn't stop analyzing it until we knew the root cause because we knew we had to understand it at every level."

BY THE END OF NOVEMBER 1966, PERSISTENT

problems with several systems on the Apollo spacecraft forced NASA to reluctantly delay the first crewed Apollo launch until February 1967. During a design certification review, issues were noted on the thrusters along with problems with the service propulsion system and the environmental control unit, and ill-fitting seats in the CM (called crew couches) required engineers from North American to travel to the Cape to enact a complicated fix.

With the first flight delayed, the rest of the flight schedule subsequently shifted. NASA hoped at least two other Apollo test flights could be sent to Earth orbit in 1967: Jim McDivitt, Dave Scott and Rusty Schweickart would test out the LM in a mission called AS-205, and then NASA picked Frank Borman, Mike Collins and Bill Anders for what they hoped would be the first flight of the Block II CM.

The news reports of the day, however, mainly focused on other critical events across the country and from the other side of the world. An uptick in US involvement in the Vietnam conflict—the bombings of the North Vietnamese capital Hanoi and the port city of Haiphong—sparked large-scale antiwar protests in cities like New York, Washington, DC, and Chicago. Civil rights protests continued against discrimination and segregation. So, while the eyes of most of the country were focused on Earth-bound problems, NASA kept its sights on reaching for the Moon.

CHAPTER 6

1967

Thinking about what we may have missed is what kept us up at night.

–GLENN ECORD, Apollo structures and materials engineer

IN MID-JANUARY 1967, FRANK HUGHES went to Downey, California, with Gus Grissom, Ed White and Roger Chaffee. North American Aviation had their own version of the simulator, called the Mission Evaluator, and it incorporated an actual Apollo Guidance Computer (AGC), so it provided different—if not better—training for the crew.

"We were about a month away from launch, and our simulator was still not working as good as it should," Hughes said. "In the Mission Evaluator, we were getting it all done, and the crew really gained a lot of confidence. They felt the mission was coming together."

The astronauts also spent considerable time in technical briefings and discussions with North American engineers. Over the past year, the crew had visited the plant frequently, watching their spacecraft come together piece by piece. They knew how everything fit together, how all the systems were designed to work. They participated in design reviews and inspections, read numerous discrepancy reports and aired complaints and frustrations. At times, they weren't certain if *Spacecraft 012* would ever be ready to fly. But now most of the open work items and hardware failures had been dealt with. The astronauts felt that while it wasn't completely perfect, their spacecraft had evolved into a very workable machine.

The prime crew of the first manned Apollo spaceflight Apollo/Saturn 204 (AS-204) inspect spacecraft equipment during a tour of North American Aviation's (NAA) Downey facility. In the foreground, left to right, are astronauts Roger B. Chaffee, Virgil I. Grissom and Edward H. White II. NAA engineers and technicians are in the background. Credit: NASA.

Left: Early morning view of Pad A, Launch Complex 39, Kennedy Space Center, showing Apollo 4 Spacecraft 017/Saturn 501, uncrewed, Earth-orbital space mission ready for launch, with a full Moon in the upper left part of the image. The 363-foot (111-m) tall Apollo/Saturn V space vehicle was launched at 7:00 am EST, November 9, 1967. Credit: NASA.

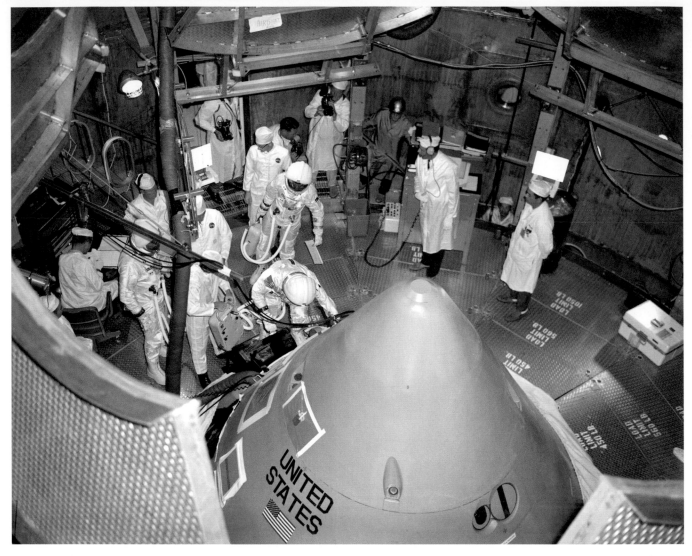

The Apollo 1 prime crewmembers prepare to enter the Command Module inside the altitude chamber at the Kennedy Space Center (KSC). Entering the hatch is astronaut Virgil "Gus" Grissom, commander; behind him is Roger Chaffee, Lunar Module pilot; standing at the left with chamber technicians is Edward White II, Command Module pilot. Credit: NASA.

Hughes learned a lot that week, too, and in the evenings he went out for dinner with the three astronauts and they enjoyed a few drinks and a lot of laughs. They had spent so much time together the past few months, and to Hughes, being with these guys was just like being with family.

Out at Cape Canaveral, Kennedy Space Center and North American technicians mated the Command and Service Modules (CSM) to the Saturn IB, and the stack was moved to Launch Complex 34. The teams worked on the list of more than a hundred significant engineering work orders and resolved technical problems. It appeared as though everything just might come together for the February 21 launch.

But a few critical tests remained, and on January 27, the astronauts needed to participate in a checkout called a plugs-out test, a full simulation of the Apollo launch countdown, overseen from both the Launch Control Center at Kennedy Space Center and Mission Control in Houston. The crew would be in the Command Module (CM), on top of the rocket on the launchpad to confirm the CSM could function properly on its own internal power. No propellants had been loaded and all the pyrotechnics were disabled, so the test was considered nonhazardous.

Grissom, White and Chaffee boarded the spacecraft shortly after 1:00 p.m. Eastern time, wearing their spacesuits and helmets so they could connect to the spacecraft's oxygen and communication systems, just like during a real launch.

From the beginning, a series of frustrating technical problems surfaced, causing holds in the countdown. When Grissom was connected to the oxygen, he reported an odor like sour buttermilk in his suit. A delay of an hour and twenty minutes ensued while technicians troubleshot the cause. The smell eventually dissipated, and finally, the pad crew sealed the spacecraft hatch, with the air in the capsule replaced with pure oxygen at 16.7 pounds per square inch, according to NASA's standard atmosphere inside all US spacecraft sitting on a launchpad.

As countdown resumed, a communications issue developed with Grissom's microphone; it couldn't be turned off. Additional problems led to frustrating periods of garbled communications and static between the crew, the Operations and Checkout Building and the Launch Complex 34 blockhouse. Various countdown functions were performed as communications permitted, but the frequent delays meant the test was running long.

Finally, an almost complete failure in communications forced another hold in the count. At 6:20 p.m., controllers announced the count would resume in ten minutes.

At 6:30 p.m., the loop crackled with static, then more garbled communications from the launch control room. Grissom said, "How are we going to get to the Moon if we can't talk between two or three buildings?"

With only static as a reply, White said, "They can't hear a thing you're saying."

"Jesus Christ," Grissom muttered, and then he repeated his query to the flight controllers, wondering how they were going to get to the Moon.

In Houston, it was 5:30 p.m. Central time. Gary Johnson, an electrical engineer, was monitoring the plugs-out test, sitting at his console in the Staff Support Room (SSR), an auxiliary room adjacent to Mission Control where experts provided technical support to the flight controllers. Johnson supported the Electrical, Environmental and Consumables Manager, and was one of the few people still in the SSR. With the test going long, most everyone from the support team had gone home for the day.

Dr. Kurt Debus, left, director of the Kennedy Space Flight Center (KSC), participated in the countdown demonstration test for the Apollo 11 mission in Firing Room 1 of the KSC control center, the same type of test the crew of Apollo 1 was participating in when fire erupted inside their spacecraft. Credit: NASA.

At 5:31, Johnson's console indicated an electrical spike, from the CM. A few seconds later, he heard some shouting on his headset, then a scream of "Fire!" and sounds of Kennedy Space Center personnel trying to communicate with the crew.

Gerry Griffin was standing in Mission Control by his Guidance Navigation and Control console. During the hold to fix the communication problems, most of the flight controllers had left the room to take a break, but Griffin remained and for some reason left his headset on. He heard some noise, like static. Then soon the word *fire* from the crew.

Guidance Officer Dutch von Ehrenfried sat nearby. "Dutch, did you hear that?" Griffin said with alarm, and then yelled to the other controllers there might be a fire on the launchpad. It took several minutes to realize the gravity of what was going on at the Cape.

The fire was inside the CM.

Back in the SSR, Johnson was still listening in on his headset to all the loops, trying to get any information he could. "Pretty soon Chris Kraft came running into the SSR and said, 'We need to play back our data so everyone can review it,'" Johnson said. "And then he told us no phone calls out of the building. We knew something crazy was going on out at the Cape, but I kept thinking since the crew were in their spacesuits, they should be okay. I was holding out hope."

Then, after several minutes, Johnson heard the Cape test conductor tell Kraft to go a private phone. Johnson's heart sank. He knew that meant the news was bad.

The news was worse than anyone could possibly imagine.

At the same moment Johnson's console indicated the spike, a wire sparked inside the spacecraft. In the pure-oxygen environment, the fire spread throughout the cabin in a matter of seconds. On the open mic, Chaffee said something that sounded like *flames*.

Two seconds later White yelled, "Hey, we've got a fire in the cockpit!" and then Chaffee shouted, "We've got a bad fire, we're burning up . . . " Then came screams. Then silence.

The communications loop at the Cape came alive, "Hey crew! Can you egress at this time, confirm? Pad leader! Get in there and help them! Gus, can you read us? Pad leader, can we get a confirmation?"

A closed-circuit TV was in the blockhouse 218 feet below the burning CM, showing a live feed of the interior of the spacecraft. Horrified ground support technicians watched the burst of flames envelop the cockpit as Ed White attempted to open the inner hatch. Technicians right outside the spacecraft in a small enclosure in the Mobile Service Structure quickly tried to open the hatch, but suddenly the spacecraft hull ruptured and a wall of fire and burning debris whooshed out, hurling the technicians backward. Other technicians on the same level as the spacecraft ran from the catwalk into the White Room, but thick black clouds of smoke billowed out, filling the White Room and two levels of Pad 34's service structure with a thick carbon monoxide haze. Some of the technicians pressed toward the spacecraft, but they began to pass out from the fumes; the next wave of rescuers from lower levels came up and grabbed the available gas masks. Still, the technicians passed out. The masks were designed for filtering out toxic fumes from propellant and were not the closed, oxygen-providing masks that were needed at this crucial moment. The remaining technicians created a plan: They formed a relay, taking gulps of air and holding their breath as long as they could to go out to the burning spacecraft and try to open the hatch.

Then a new danger became evident. Someone in the Control Center feared the CM might explode, or the fire might set off the launch-escape system atop the entire spacecraft stack. Either event could ignite the entire service structure. Some technicians left while they could, but others remained, helping the injured, trying to rescue the astronauts.

Roughly five minutes after the fire started, the final relay of technicians opened the searing hatch. The fire had quenched itself when atmospheric air rushed into the CM through the ruptured hull.

The astronauts were dead. The tubes connecting their spacesuits to the oxygen melted in the extreme heat, and the astronauts were asphyxiated from toxic fumes, overcome from the flames and heat. The time from the first indication of the fire to the final crew communication and loss of all telemetry was seventeen seconds.

FRANK HUGHES WAS ON-SITE AT THE SIM

facility at Kennedy Space Center that evening, working with one of the contractors when, at about 6:40 p.m., someone stuck their head through the door and said, "Something wrong with the crew? Have you heard anything?"

"No, I don't know anything," Hughes said. "They're out in the spacecraft." Hughes went in his office to call his boss, but the phone was already ringing. From then on, all hell broke loose. The first caller said there had been an accident, the next one said the crew had been injured. Each time the phone rang the news got worse and worse and worse. The crew was dead.

Hughes was so young, he'd never experienced death before. No one he had ever known in his entire life had died; and now three of his friends were all dead, in the same day, in the same horrible way.

Hughes's branch chief, Riley McCafferty, came in the office just as the phone rang again. It was Deke Slayton.

"We need to get somebody to go out and check the switches in the spacecraft," Slayton told McCafferty. "They're not sure that the crew might've thrown a switch or something that caused it. Can you get one of your guys to head out?"

"Frank's here," he said. Hughes followed McCafferty out to his car and they drove over to Launchpad 34. McCafferty explained that they wanted Hughes to slide in the spacecraft on a board and write down all the positions of the switches before they took the bodies out.

The idea of actually going into the spacecraft . . . Hughes couldn't even think about it. It was all a shock. These three guys he had worked with, horsed around with, been out to lunch or dinner with almost every day. It was like he'd just lost three older brothers. There was guilt. What had they done wrong? Could he have done something to keep this from happening? What the hell had they been doing? What had anyone been doing? He couldn't even think. But, Hughes decided, he was going into that spacecraft and he'd just do his job, because he wanted to help Gus, Ed and Roger and he wanted to help everyone figure out what the hell had happened.

Hughes and McCafferty went up to the spacecraft.

Chaffee was still in his couch; he had undone his harness. White was still in the middle seat, twisted and reaching up, trying to open the hatch. Grissom had rolled to one side, trying to get away from the worst of the flames, nearly on top of White. Their spacesuits were charred, melted together. The interior of the CM looked like someone had used a blowtorch, with some areas totally destroyed while adjacent areas were virtually unscathed.

Hughes looked around. Firefighters and medical teams tended to the twenty-seven technicians who were overcome by the carbon monoxide fumes and heat; some were burned.

Slayton came over, clearly distraught. He looked at Hughes and said, "You don't have to go in there, Frank. Some of the switches are melted; they're going to have to take them all apart to find out what happened." Then he paused and added, "They didn't cause this. They didn't."

Later, Slayton followed Hughes and McCafferty down the elevator from the gantry and then said, "Since you're here, Frank, go over to the Operations and Checkout Building. People will be coming in from all over the country, they're going to start an investigation right away. We can put everybody up in the crew quarters. Check them in and get them whatever they need."

Astronaut Frank Borman got there first; he had jumped in his T-38 and flew directly from Houston. Then other astronauts and NASA officials arrived, getting there as soon as they could. Hughes stayed all night as the quartermaster, ensuring everyone who needed a room got one. He ordered extra food and found extra cooks; he made sure everyone had breakfast in the morning. He was busy until 10:00 a.m. He didn't even think twice about it—he was glad to help any way he could. And being busy meant he didn't have any time to think. It had been a horrible, horrible night.

The charred remains of the Apollo 1 cabin interior. Credit: NASA.

THE NIGHT OF JANUARY 27 WAS SUPPOSED to be a celebration. That day in Washington, DC, NASA hosted an Apollo Executives' Conference, a congratulatory affair for the contractors who contributed to Gemini and a welcome party for the Apollo contractors. Several attendees of the conference were invited to the White House to witness the signing of the Outer Space Treaty, a UN document that stated no nation could claim any region of space. President Johnson described the treaty as "the first firm step toward keeping outer space free forever from the implements of war." A black-tie dinner had been arranged with James Webb, George Mueller and other top NASA officials and corporate leaders.

Robert Seamans had already arranged for a small dinner party at his house with Doc Draper and a few other close associates, so Webb had told him to go ahead with his own plans. When Seamans reached home, his wife, Eugenia, was talking on the phone. "He's coming in the door now," she said, handing him the receiver.

George Low was on the line. "They're all three dead," he said.

"What three?" Seamans asked.

"The three astronauts."

Devastated, Seamans returned to his office at NASA Headquarters and started making phone calls.

A test of the Apollo Command Module, during postfire testing at the Manned Spacecraft Center in Houston. Credit: NASA.

"For many months, we would conduct flammability tests, outfitting the interior of a pressurized spacecraft with the new materials," Chaffee said, "and then put an ignition source inside it and initiate a fire. It was all heavily instrumented and filmed where we could watch the progression of the fire, measuring everything from the temperatures, smoke patterns and chemical content."

Besides changes in the spacecraft, the Apollo management structure transformed as well. George Low replaced Joe Shea as Apollo spacecraft program manager in Houston; in Downey, Harrison Storms was out, and a proposed merger

between North American and aerospace company Rockwell-Standard took place in April, forming North American Rockwell. Under new leadership, the company assigned a spacecraft manager with a personalized team to each vehicle and added a program manager whose only assignment was addressing safety.

While the tragedy of the fire could have torn NASA apart with infighting and accusations, instead the agency seemed to pull together as never before. Two people in particular had an important impact: George Low and Frank Borman.

"George just brought us together," said Glynn Lunney. "He said, 'Okay, we have a problem but now we have to recover and get this program back on track.' Through his intellect, his energy and his way with people, George was able to affect a 'recombining' wherever we may have been fractured. He let us have our guilt trip for a while, but then he said, 'Okay, now let's go do this,' and he melded us into a real force."

Low created a special "Change Board," where all changes to the Apollo spacecraft necessary to make it flightworthy would need to be approved before implementation.

Borman led a "tiger team" at Downey—with authority to make on-the-spot decisions on the hatch, wiring configurations and other improvements that had been planned even before the accident. As North American Rockwell engineers went over the spacecraft piece by piece, the team lent assistance when necessary. But somehow with his stubborn single-mindedness, Borman found a way to rally the troops.

"Although Frank's job was primarily technical," Lunney said, "I think the people there had emotional wounds of sorts to heal to get themselves back on track. And I think Frank—God bless him—brought a lot of that to the Downey group and helped with the healing process, moving them on to the stage of transformation, of moving forward. He said, 'Okay, we've had this problem, we've had our recriminations. Now let's get the hell on with it.' He made them feel they could push through this and really move forward."

Across NASA, the trauma of the fire precipitated renewed dedication and cooperation among all the NASA centers and the Apollo contractors. NASA leadership provided greater stability, influence and control. While in some areas at NASA the postfire recovery turned into a frantically busy period of enacting the required fixes, in other areas the pause in racing toward the launchpad provided time for improvements and reflection. It gave engineers the time to do what they did best: to think of all the what-ifs that might happen and how they could be prevented.

"I think throughout NASA, we had felt bulletproof in a way," Norman Chaffee said. "We'd experienced a lot of success, and maybe our thinking had become a little skewed, since if this thing or that thing hasn't bitten us so far, it must be okay. The Apollo 1 fire really brought home that we can't tolerate any waiving of requirements or cutting of corners, and we needed to do our job right. It brought all of us to stand up and say, 'This is my personal job to make sure that everything is right.'"

Astronaut Frank Borman looks over the Gemini 7 spacecraft during weight and balance tests. The tests are conducted in the Pyrotechnic Installation Building, Merrill Island, Kennedy Space Center, as part of preflight preparation. Credit: NASA.

WHILE THE MAJORITY OF MSC WORKED feverishly on testing and acquiring new materials and procedures after the fire, the sim team's schedule slowed.

"All the engineering guys were off redesigning the Command Module and fixing all sorts of things," said Jay Honeycutt, who joined the sim team in 1966. "But twice a week, we'd all meet in the Control Center and run Apollo 7 sims. Schirra and his crew would be down at the Cape in the simulators, Glynn Lunney was the flight director. Every Tuesday and every Thursday we'd crank everything up."

Simulators at the Manned Spacecraft Center. Credit: NASA.

The repetitive nature of the work during that time helped everyone in the Sim Group get into a certain rhythm, and they also learned how to do their business on a larger scale. In every aspect, Apollo was so much bigger than what they'd done in Gemini, with multiple spacecraft traveling farther and doing more. The simulators and Mission Control were now more sophisticated, with several computers now handling numerous data streams.

The sim team spent time writing the "scripts" of their upcoming simulations, coming up with realistic mission scenarios and plausible problems but throwing in some occasional fun as well. They studied schematics, talked with engineers and sat in on mission-planning meetings, figuring out how to best point out the weaknesses and strengths of the flight control team, the astronauts and the spacecraft. Harold Miller told his team that the SimSup held the final say on how well the team performed, and the team knew their research on possible faults or failures might hold a vital key to mission success.

The sun peeks through what remains of the gantry on Launch Pad 34 in 2017 at Cape Canaveral Air Force Station in Florida where Virgil Grissom, Edward White and Roger Chaffee lost their lives. Credit: NASA/Ben Smegelsky.

AMID THE REPERCUSSIONS OF THE FIRE

and during the rebuilding period, public support for NASA remained relatively strong. The prevailing sentiment seemed to be that the three astronauts should not die in vain, and NASA should push on with the missions to the Moon in honor of Grissom, White and Chaffee. Fuel for that sentiment came from an article published by the Associated Press shortly after the fire, with a stirring quote attributed to Grissom: "If we die, we want people to accept it. We're in a risky business, and we hope that if anything happens to us it will not delay the program. The conquest of space is worth the risk of life."

In going forward, the widows of the Apollo 1 crew—Betty Grissom, Pat White and Martha Chaffee—had requested that NASA officially name the mission their husbands never got to fly as Apollo 1, and also asked that mission designation be retired. NASA agreed, and then went on to change the way remaining flights were named. There were two test flights, originally named AS-202 and AS-203, that were canceled and weren't renumbered in the Apollo series, but subsequent missions would be named beginning with Apollo 4. Apollo 4 was a test flight, scheduled for November 1967, and it helped NASA get its groove back.

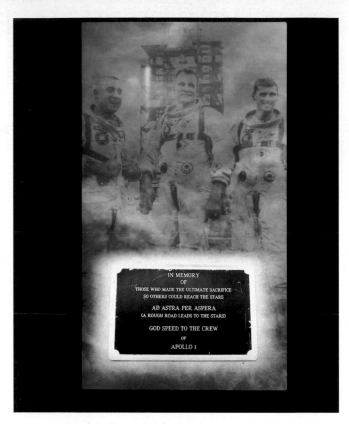

A display screen at Kennedy Space Center in 2017 showing the memorial plaque that is in place at Launch Complex 34 in a tribute to the crew of Apollo 1 who perished in a fire at the launch pad on January 27, 1967, during training for the mission. Credit: NASA.

DESPITE THEIR EARLIER RESERVATIONS

about conducting all-up testing, Wernher von Braun and his team were ready for the first flight of their Saturn V rocket without their customary gradual series of test flights. On November 9, 1967, everyone at Kennedy Space Center and the surrounding area witnessed the grandeur of the largest rocket ever flown. While earlier flights of the Saturn 1B had impressively blazed a trail and provided a hint of what a powerful rocket launch could be like, there was no way to prepare for the incredible sound and ferocity of the monstrous 363-foot (111-m) tall Saturn V. As it powered away from the new Launch Complex 39, the ground and buildings shook for miles around. Dust and debris fell from the ceiling of the newly built Launch Control Center. Inside the new CBS News building on-site at Kennedy Space Center, commentator Walter Cronkite shared the experience with viewers around the country.

"My God, our building's shaking here," he said, his voice trembling from the rocket's vibrations as well as emotion. "Oh, it's terrific, the building's shaking! This big blast window is shaking! We're holding it with our hands! Look at that rocket go into the clouds at 3,000 feet! Oh, the roar is terrific!"

NASA was counting on this rocket to allow future flights to get to the Moon, and its maiden voyage was successful by every measure: from its majestic and slow rise from the launchpad to the third stage boosting a test version of an Apollo CSM into orbit. After eight hours and thirty-six minutes of flight, the CM splashed down in the Pacific Ocean.

After the Apollo 1 fire, many at NASA had adopted the phrase "ad astra per aspera," meaning "a rough road leads to the stars." Later, that phrase would be placed on a plaque on Launch Complex 34 after it was decommissioned, along with another inscription that reads, "They gave their lives in service to their country in the ongoing exploration of humankind's final frontier. Remember them not for how they died but for those ideals for which they lived."

The road to the Moon had certainly been rough in 1967, but the Apollo 4 launch showed that reaching this monumental goal just might be possible. The giant rocket had provided a stirring and triumphant lift in spirits that everyone needed.

Right: The Apollo 4 space mission launched from Pad A, Launch Complex 39 at 7:00 a.m. on November 9, 1967. Credit: NASA.

CHAPTER 7

1968

You couldn't have blown us out of there with a stick of dynamite. Everyone was having so much fun.

— JAY HONEYCUTT, Apollo simulation team

A MASSIVE HURRICANE RAVAGING THE Houston area. Or Cold War tensions escalating and the Soviet Union sabotaging NASA's Manned Spacecraft Center (MSC) facilities. What would happen if, somehow, the Mission Control Center had been destroyed or incapacitated and an Apollo crew was on their way home from the Moon, desperately trying to contact the ground?

"It was one of those what-if scenarios that many of us had never even considered," said Johnny Cools, who worked in flight control in MSC's Real Time Computer Complex (RTCC). In the aftermath of the Apollo 1 fire (on January 27, 1967), program managers took stock of all the possible things that could go wrong during a mission. One idea that came up was the concept of an off-site, backup Mission Control, just in case a catastrophic event occurred in Houston.

Within the Flight Support Division, Cools was tasked with preparing the procedures and conducting a test run of an Emergency Mission Control Center (EMCC), set up at the Goddard Space Flight Center in Maryland. In early 1968, several imaginary disaster scenarios were considered as Cools, Retrofire Officer Chuck Deiterich and a few other

The Real Time Computer Complex at the Manned Spacecraft Center in Houston. Credit: NASA.

flight controllers and IBM computer programmers grabbed what was called a checkpoint tape from the RTCC, carried it on board an aircraft and made their way to Goddard to test operational control of a simulated mission from the EMCC.

Left: The Apollo Command and Service Modules inside MSC's Space Environment Simulation Laboratory's Chamber A, for a test in 1968. The light banks simulate light from the sun. Credit: NASA

MSC Mission Control Center, Building 30 in 1964. Credit: NASA.

"For every mission, we had the standard procedure to create checkpoint tapes or restart tapes every hour and a half or so," said Cools, "enabling us to have a backup 'snapshot' of the data and telemetry from the flight should we lose the main computer, the Mission Operations Computer, for whatever reason. So for this test, we hand-carried one of those magnetic tape reels, flew to Goddard and performed an initial program load of the mission software to the computer at the EMCC and got everything operational. That test assured us we had the ability—should the need ever arise—to compute a return trajectory for the crew and help them return safely to Earth."

Potential catastrophes aside, the MSC's computer center had other backup procedures as well, but the RTCC was the heart and soul—and the brains—of Mission Control's real-time data-processing abilities. During the stand-down after the Apollo 1 accident, the RTCC was refurbished and the previous IBM 7094 computers were replaced by five brand-new IBM 360-75Js. In total, the new computers had a storage capacity of about 1 megabyte. During each flight, one computer would be assigned as the Mission Operations Computer (MOC). The MOC handled the real-time data: the telemetry coming

in from the spacecraft, the commands going out to the crew, radar processing and trajectory and ephemeris planning. Another computer was designated as a backup, called a dynamic standby, and it performed the same operations as the other flight control computer; however, it was considered an emergency backup system, but it could be brought "online" at the push of a button, becoming the MOC within a second or two. One of the other computers was used by the simulation team, and the remaining computers were used for other computing needs at MSC.

"During the Apollo flights, we would have about nine tape drives that were recording all the data from the mission," said Cools. "We always had backups because we wanted to make sure that if any drive had a hardware failure that it automatically switched to the next drive and we wouldn't lose any data."

The RTCC was located on the first floor of the Mission Control Center, and as it processed all the data from the mission, it provided the ability for data to be displayed on the cathode ray tube (CRT) displays on the flight controller consoles, on the large front display screens and on other data-output devices throughout the Mission Control Center.

Those who worked in the RTCC weren't just computer experts—they were flight controllers who were responsible for the operation and control of the MOC's software programs as well as for interfacing with the flight control team in Mission Control.

"We were assigned to sit alongside the flight controllers on consoles during the simulations and testing of operations," said Cools, "and then we would write the requirements from each flight controller and then hand them off to IBM to be programmed. Once the programming was done, we'd check the systems and the processing to make sure everything was working with internal testing before handing it off to flight control for simulations and then flight operations. We were literally a thread running from the very beginnings of figuring out how this was all going to be done, down to sitting at the consoles and supporting the flights."

There were five controller positions for the RTCC, and each was responsible for providing assistance to a number of different Mission Control flight controllers. They needed to be available to respond to any requests for specific data, as well as requests from the backroom support teams.

"The MOC would complete the requested processing with the results displayed on the CRTs," Cools said. "For example, computing a new trajectory or ephemeris based on the latest radar vector. Some of the data processing was automatic, such as processing all telemetry parameters, which was done once every second. It was our job to see that all the data was processed properly and the output was displayed correctly."

In short, without the RTCC, NASA's flight controllers would be flying blind. After the computer upgrade, and with all the testing and simulations, the RTCC and flight control management teams knew they were ready for the first crewed flights to get under way and could give a hearty "Go!" for Apollo.

BY LATE 1967 AND INTO 1968, IT WAS

common for several astronauts at a time to be out at North American's facility in Downey, California. North American had accommodations for about twenty people to stay in private rooms. The company also provided cars for the astronauts: Ford Mustangs.

"If they were on facility," said Rich Manley, "they were assigned a Mustang, and they could pick it up at LAX. They could come and go as they needed, and they would be on site for four to five days at a time, maybe even weekends. We had seven days a week testing and the astronauts were part of it. The primary crew and backup crew were each assigned to the spacecraft they'd be flying, and it was their responsibility to follow their spacecraft along its development, monitor it and to get to know it well."

Each astronaut was assigned to specialize in a specific system, to know it inside and out, then follow the technical development and mainline testing—many times adding input on the design or testing. They'd have meetings with the other astronauts to share details of changes or anomalies, just to keep each other informed on the status of the spacecraft. The astronauts regularly came into Manley's lab in the engineering ground station.

"We'd have the coffee pot on, and they would come in and BS sometimes, but mostly it was strictly business," he said. "They were all very precise and if you were not on board ready to go when the switch turned on to do the test, you better damn well be there, they'd have you thrown out."

But Manley also found the astronauts were either 100 percent work or 100 percent play, and when it came to play, interesting things could take place.

One night when Manley finished work at about midnight, he drove out of the parking lot, about to turn right on Lakewood Boulevard. Lakewood ran northeast–southwest in front of the North American plant, with three lanes in each direction. In the late '60s, there was a mile-long stretch where no other streets came onto the highway, except for the access road to the plant.

Before turning, Manley noticed two sets of headlights barreling down the highway, bat out of hell, one car right on the butt-end of the other. As the cars flew past, Manley saw they were both Mustangs. Attached to the trunk of the lead car was an Apollo docking port, and on the hood of the second was a docking probe.

"They were docking and undocking with each other, probably doing about 50 to 60 miles per hour down the highway," said Manley, who almost couldn't believe what he was seeing. "I never found out who it was because when I mentioned it the next day, they all said they didn't know anything about it."

AFTER THE APOLLO 1 FIRE, WHEN NORTH American Rockwell changed the designs for the Block II version of the Command Module (CM), some of Honeywell's designs required modifications as well. And the Instrumentation Lab's upgrade to their own Block II version of the Apollo Guidance Computer (AGC) affected the work that Earle Kyle and his colleagues were doing too.

The requirements set by the Apollo program were many. Every part needed to include an individual stamped identification so that if there were any problems, the part could be traced back to its origins on the assembly line. Kyle found that particular requirement quite useful several times while tracking down a specific issue. But the paperwork requirement was seemingly endless, with no tangible benefit.

"We had so much paperwork in documenting everything," Kyle said, "that we used to joke NASA had a secret plan that if the Saturn V didn't work, they could climb to the moon on the stacks of paperwork."

Kyle and his colleagues always wondered if anyone at NASA or North American Rockwell had time to read all the paperwork they generated. If they added up all the paperwork they spit out, plus the paperwork from the other contractors, reading it would be humanly impossible. After compiling one particularly laborious and lengthy 3-inch (8-cm)-thick report, Kyle and his coworkers decided to conduct a test. Behind some well-placed introductory pages, they put a Minneapolis phone book inside the binder. No one ever called them out on their prank, confirming Kyle's assumptions.

Honeywell's design lab included a big blueprint machine that could print out the schematics of their electronic designs on huge rolls of paper. The engineers would tack the large blueprints on the wall, all the way around the room and they could walk around and make notations. If they realized a problem with the design, they'd rip down the paper and start over.

Kyle and his colleagues constantly used the Honeywell Visicorder, a multichannel strip chart recorder (similar to an EKG machine) that could test out the electronics and provide a printout of how each chip or diode was functioning. Standing about 3½ feet (1 m) high, the Visicorder was on wheels so it could be moved wherever it was needed. The thermal paper for the printouts came from another Minneapolis company, 3M, and the 11-inch (28-cm)-wide paper would continuously stream out, displaying the squiggly lines that showed details of the performance of a component. But paper jams happened frequently.

"I can remember many, many nights at 3:00 a.m., lying on my back on the cold concrete floor in the lab where there was no heat in the winter," said Kyle, "reaching up into that machine to unjam the paper roll. I'd be so exhausted I'd just lay on the floor with the warm thermal paper falling down on me. I'd take a quick nap, and then go back to analyzing the printout to see where things went wrong."

Each of the electronic modules Honeywell built had to be tested and subjected to the torture they might endure during a launch, while in space or being stored for years in the warm, humid Florida weather. Kyle and his colleagues used vibration tables, vacuum chambers and environmental simulators that included salt spray and temperature extremes. They also had to make sure that the big specialized computer they built, the Bench Maintenance Equipment (BME), could operate after sitting in the Cape's humid, salty air. (The BME in Minneapolis didn't have that problem as it had its own glass-walled, air-conditioned room.)

Kyle was responsible for the analog computer that fed signals to the display for the Flight Director Attitude Indicator (FDAI), the eight ball. The display told the crew which way the engine on the Service Module (SM) was pointing when they needed to make in-flight course corrections. During the first several "builds" of this piece of flight hardware, the modules worked well. But during testing of the sixth set, some electronic signals started to drift—they would intermittently be too close to being outside the required voltage or frequency range. The drift could be a problem because Minneapolis-Honeywell needed to ensure their modules would work correctly two or three years down the road. Some of the modules they were building wouldn't be flown until a later Apollo flight.

"There are two kinds of problems you run into in this business," said Kyle. "One is completely broken, which is usually easy to find and fix. But intermittent problems, those are hard. Sometimes it can take weeks or months to find the culprit."

The Honeywell Stabilization and Control System that operated the thrusters on the CM and the LM used a hybrid of both analog and digital silicon computing chips that were packaged in small metal rectangular "cans." These cans were about ½ inch (1 cm) long and about ¼ inch (0.5 cm) wide with wire leads coming out on both sides so they looked like daddy longlegs. These leads were welded onto circuit boards with other external components like power resistors and capacitors. Several circuit boards were then placed in various special sealed shoebox-size boxes, or modules, containing hundreds of circuits. So finding the culprit responsible for the problem of signal drift was painstakingly difficult. It became a months-long investigation, and since Kyle had a reputation as a problem solver, he was appointed to lead a team to figure it out.

"My boss told me I had unlimited resources, but I needed to solve the problem in less than three months or we could lose the contract," Kyle said. "This was one of the biggest contracts Honeywell ever had, and so I eventually had a team of eighty people working on it, and we used all sorts of exotic gear, like electron microscopes, wide bandwidth oscilloscopes and X-ray machines, to run our tests to figure out this spooky, intermittent problem."

Finally, Kyle and his team were able to determine the problem: It came from the advances made in the mid-1960s with new miniaturization techniques in building computer circuits. "The supplier had figured out a way to make the microchip inside the can smaller, but they didn't tell us about it," said Kyle, "and we couldn't know because the can surrounding the chip was still the same size with the same part number on it. When Honeywell's specialized technicians used their capacitive discharge welding machines to fasten the microchip cans onto the circuit boards, the smaller chips inside each can were heat stressed where the larger previous designs weren't. And this set the stage for the hardest type of problem to solve—one where you don't have a catastrophic part failure, but one in which the part is stressed and is now going to be flaky and unreliable and go off on its own from time to time and do all sorts of strange things that are hard to find the reason and the cause and then, of course, the fix."

But still, Kyle marveled how this era seemed to be a convergence of time where the digital age met the rocket age, where everything came together to make Apollo possible.

"And the same thing was true for me too," Kyle said. "I seemed to run into the right people at the right time to solve a problem or to be exposed to an idea or experience that helped me along the way. I had so many serendipitous things in life, it was spooky how things just meshed together. I always said, you couldn't make this shit up in a million years."

Of course, with the pressure of meeting the goals and timelines of Apollo, there were conflicts and people who made life difficult. One of Kyle's bosses was a cigar-chomping tyrant, and when he walked through the maze of cubicles and heard someone laughing, he'd threaten to fire them. Another was a Bible-thumping zealot, continually grumbling about the waste of money on Apollo when there were so many problems here on Earth.

One meeting, in particular, stands out during that time. Kyle and his team were meeting with members of other engineering groups at Minneapolis-Honeywell, trying to sort out some issues that had been plaguing the various groups. Each week the groups met in design-review committees to make sure they had considered everything in ensuring that nothing like the Apollo 1 fire would happen again. Among one of the groups was a black engineer, but he had to leave the meeting early. A short time later, after the group worked through a particularly tricky problem by identifying a subtle engineering error that slipped by the design review, Kyle's boss said, "Let's be more careful. I don't want to see any more 'nigger in a woodpile' incidents like this again, as we can't keep falling behind schedule."

When Kyle heard that word, he sharply drew in his breath and his blood started to boil. His initial instinct was to call his boss out for his derogatory insensitivity, his bigoted racial views. But he knew a private conversation would better serve the situation.

"I was hot because I'm half black but very light skinned so obviously he didn't know he offended anyone who was left in the room," Kyle recalled. "Clearly he waited until the dark-skinned engineer left before making his racist comment. I didn't want to embarrass him in front of the others so waited until later to explain he should never say that again while I'm around or I'd probably 'deck him.' Of course, he turned beet red, apologized all over the place, said it would never happen again. He probably wouldn't have said it if he had met my very dark-skinned wife. He'd probably think we were just an interracial couple."

Kyle had experienced this before, had "been to this movie many times before" in his life.

When he was in high school, he overheard his trigonometry teacher talking to the track coach. He said, "I wish I could figure out a way to not give that Kyle kid an A grade for this course as I'll be damned if I'll help a nigger be the valedictorian of Minneapolis Central High School."

The coach was confused. He thought Kyle was white. The math teacher explained, "I've had his older brother and sister and they are very dark, and so is his dad."

But Kyle received the A in trigonometry and became the first African American valedictorian of Central High. In his valedictory speech, he talked about his dream of sending people into space. And now, the convergence of all the right elements in his life meant he could work to make that dream a reality.

AT 7:00 A.M. ON MAY 6, 1968, NEIL

Armstrong took off to conduct simulated lunar landings using the Lunar Landing Training Vehicle (LLTV). Armstrong had been assigned as backup commander for the Apollo 9 mission, and all prime and backup Apollo commanders were expected to complete training in this awkward-looking "flying bedstead," as the astronauts called it. The LLTV's tendency to be persnickety in operation was superseded by the realistic training it provided for flying the Lunar Module (LM). To avoid any extra instability, standard procedure was to fly it early in the morning at Ellington Air Force Base to avoid the Gulf breezes that usually came up later in the day.

After conducting landing maneuvers for about five minutes, Armstrong rose to about 200 feet (61 m) above the ground when he suddenly lost control of the vehicle. He quickly made the decision to eject just seconds before the LLTV spun out of control and crashed onto the runway in a spectacular fireball. Incredibly, Armstrong's parachute opened and filled, even at such low altitude. He floated down safely to the nearby grass and was uninjured except for biting his tongue when he hit the ground. Firetrucks standing by at Ellington quickly came to the rescue and put out the flames, but the LLTV was a total loss. Anyone who saw the event couldn't believe Armstrong escaped unscathed. It all happened in just seconds and was an incredibly narrow escape.

Word about the accident slowly made its way around MSC, and later that morning Glynn Lunney came by to talk with astronaut Alan Bean, who shared an office with Armstrong. As Lunney walked into their office, the two astronauts were looking in a textbook, talking about lunar surface features.

Lunney said, "Hey, that was a pretty close call this morning, Neil."

Bean looked at both of them quizzically. "What are you talking about?"

Armstrong hadn't mentioned his brush with death to Bean, even though the two were good friends and had been in the same room for a couple of hours. Bean later said, "I can't think of another person, let alone another astronaut, who would have just gone back to his office after ejecting a fraction of a second before getting killed."

Neil Armstrong parachuting to safety after ejecting from the LLTV-1, which is seen burning on the ground after it crashed on May 6, 1968. Credit: NASA.

Years later, Ken Young from the rendezvous team was chatting with Armstrong at an MSC reunion event and he took the opportunity to ask about the LLTV crash. "You know, Neil, I heard from Glynn that you never even told Bean-O that you almost bought the farm that day when you bailed out of the flying bedstead," Young said. "Is that really true?"

Armstrong smiled and looked down bashfully. "Well, yeah, I suppose that's true," he said slowly, and then looked up at Young with a wry grin. "But do you know what the worst thing was about that day?"

"That was a pretty bad day, Neil," Young said. "What could be worse than almost getting killed?"

"Well," said Armstrong, "when the LLTV crashed, I floated down into that tall johnsongrass alongside the runway, and the parachute dragged me through the grass. I ended up getting chiggers in both legs, and it took more than a month to get rid of 'em."

Left: Astronaut Neil Armstrong flying the Lunar Landing Training Vehicle (LLTV) at Ellington Air Force Base in Houston, Texas. Credit: NASA.

Apollo Command and Service Modules (CSM) being moved within Kennedy Space Center's (KSC) Manned Spacecraft Operations Building in 1969. Credit: NASA.

Technicians prepare a Lunar Module for the LTA-8 test in the vacuum chamber in the Space Environment Simulation Laboratory at the Manned Spacecraft Center in 1968. Credit: NASA.

THE NEW BLOCK II VERSION OF THE

Command and Service Modules (CSM) was undergoing design and engineering reviews at both North American Rockwell and within NASA to certify the designs for flight-worthiness and flight safety. Certification meant plans could go ahead for scheduling the Apollo 7 and Apollo 8 missions. But before the designs could be verified—and before any humans could fly in these spacecraft—engineers needed to conduct critical tests in the enormous vacuum chamber at MSC, the Space Environment Simulation Laboratory (SESL).

"The management asked me to lead up a team of several hundred people to do what we called the LTA-8 and 2TV-1 tests in our big space chambers," said Bob Wren. "We wanted to put the spacecraft and astronauts through simulated flights lasting several days to make sure everything was going to work in the vacuum and temperatures of space. It took us several months to even set these tests up, and we even brought in the designers and technicians from North American and Grumman to help us."

NASA considered both these tests as "mission constraints" for the flights of Apollo 7 and Apollo 8, meaning if anything went wrong or problems were uncovered the flight schedule would need to be delayed until the issues were resolved.

The first test, LTA-8, tested the LM, which was of great interest, because windows in the LM had shattered during a cabin pressurization test in December 1967. The LTA-8, a Lunar Test Article—a nonflying test version of the LM—was placed in the smaller vacuum chamber B to undergo four separate tests, from between one and ten days in early May to the beginning of June. Besides the vacuum environment inside the chamber, a bank of specialized lights simulated the light and heat from the sun, while temperatures could also be regulated to the −250°F (−157°C) cold of space. The simulated crew of astronauts, Jim Irwin and Gerald Gibbons, entered the LM in the chamber only for short periods and conducted simulated separation and docking maneuvers. All systems worked well, and the crew experienced only minor difficulties during ingress and egress of the chamber.

The public expressed high interest in the chamber tests, which were covered extensively in the media. NASA issued press releases on each test. With no "disabling anomalies" uncovered, James McLane triumphantly announced, "The craft passed its final preflight tests with flying colors. We have removed the last uncertainty. We're home free!"

The Apollo Command and Service Modules inside Chamber A of the Space Environment Simulation Laboratory at MSC for the 2TV-1 test. Credit: NASA.

The 2TV-1 test for the CM garnered even greater interest because it would be conducted in the enormous Chamber A, and the crew would remain in the chamber for eight days straight. The test for 2TV-1 (which, in NASA's acronym-heavy vernacular, stood for Apollo Block II thermal-vacuum) relied on expertise garnered during tests of the Block I CM designs conducted in 1966. This problematic eight-day test with three astronauts included a malfunction of the vacuum depressurization system, problems with the light banks, ice buildup on the heat shield and excess moisture inside the spacecraft. The biggest threat to the continuity of the test came when the urine dump line froze, meaning urine needed to be stored in bags for the duration of the test. The strangest issue uncovered in the 1966 chamber test came from problems with the specialized underwear worn by the crew. The underwear had been designed especially for space travel, but the crew was forced to take the underwear off and stow it securely after it was discovered the material off-gassed poisonous lithium fluorine gas.

But with fixes made and lessons learned, Wren organized a crew of about seven hundred people in the spring of 1968 for the 2TV-1 test to ensure that the CM maintained the proper environment for crew and equipment in the extreme vacuum and temperature of space. With a rotating base, the spacecraft could be turned in the "barbecue mode" to test the passive cooling technique.

First, Wren's team conducted a three-day test without the crew inside to verify the integrity of the vehicle and the systems. Then, on June 16, 1968, astronauts Joe Kerwin, Vance Brand and Joe Engle entered the CM inside the vacuum chamber. During the next eight days, the crew performed many of the functions as if on an actual spaceflight, including eating and sleeping.

"We hooked up communications and data systems to Mission Control," said Wren, "with cables running through the underground tunnels, so the crew operated everything as if they were in flight."

Interior views of the Command Module and astronaut Joe Engle during the seven-day manned thermal vacuum test, 2TV-1 in chamber of Building 32 at MSC. Credit: NASA.

The consoles for operating and monitoring the vacuum chambers in the Space Environment Simulation Laboratory at MSC. Credit: NASA.

The crew tested guidance and navigation equipment, activated and checked out spacecraft systems and simulated engine firings. Meanwhile, chamber operators put the spacecraft through several phases of flight, simulating various temperatures and lighting conditions. Technicians and engineers worked around the clock in twelve-hour shifts, monitoring systems and activities inside the chamber with television monitors and a myriad of control panels.

But all went well. "Those eight days were incredibly fun," Wren said, "and the biggest problem came from condensation, with everything dripping inside the CM from all the human exhaust." Engineers instituted subsequent changes to the environmental controls in the real CM.

Kerwin felt the biggest test came when assessing the new fireproof materials, breaking ground to enable the flight of Apollo 7, which would soon be scheduled for early October 1968.

But eight days confined in a vacuum chamber had its challenges.

"We didn't think of it as being quite such a monumental thing," said Engle. "We were kind of bored in there, actually. I think I pulled on my hunting and camping skills to live in a confined area. Being confined in a tent while it's raining for several days with a couple of guys—that was good training for 2TV-1."

If the crew was bored, imagine sitting at a console for twelve hours during a sleep shift.

"One night, one of these technicians got up from his station and told the guy next to him, 'I'm going to the bathroom,'" McLane laughed. "He was never seen again. He'd just had enough of it."

Another evening during the test, Apollo program manager George Low visited the facility and noticed a technician paging through a copy of *Playboy*.

"You let your guys do that all the time?" Low asked McLane.

"Look," he replied, "we let them do anything when they're on a duty station like this, and they might need some stimulus to stay awake."

WITH THE FIRST CREWED APOLLO FLIGHTS

on the horizon, the simulation schedule ramped up. And soon, the Simulation Branch developed the reputation of being diabolical. But that wasn't their intention at all.

"We didn't do 'gotcha' sims," said Jay Honeycutt. "It wasn't our goal to make anybody look bad. Our job was to develop a team that could work smoothly together."

By 1968, the sim team consisted of about forty people, some specialized in the CSM, others in the LM or trajectory.

"We would divide ourselves up into those positions, but we really were only one team," Honeycutt said. "It didn't take forty people to run a sim, but it took about ten, and there were five flight control teams. So, they had us outnumbered."

No one had time to keep good notes or records, and some sim scripts were written on the back of bar napkins while others were jotted down on paper, only to be misplaced later.

The sims could be a big endeavor, encompassing as many as two hundred people. The ten console positions in Mission Control each had several people in the various backrooms for support, such as the Mission Evaluation Room (MER) and the Staff Support Room (SSR). The SimSup couldn't give everyone something to do at once because of the flight control team's pyramid structure. "You couldn't blow up every console with enough work to keep everyone busy," Honeycutt said, "because the flight director would go under just from the sheer magnitude of everyone wanting to talk to him at one time."

The sim team learned how to listen to several conversations at once. Every console position in Mission Control had a voice loop with their back room, and every position could talk to the flight director and hear the crew. The SimSup could punch into a variety of the voice loops and listen to the conversations in order to modulate how all the problems were dealt with. Sometimes the problems might come from a simulator glitch and they'd have to reset and start over.

Some simulation runs lasted the entire day, from 7:00 a.m. to 8:00 p.m., if they were practicing going to the Moon or being in lunar orbit. For shorter events like launches or landings, they would run 8 to 12 cases a day.

Astronaut Michael Collins, Command Module pilot of the Apollo 11 flight, is seen inside an Apollo Command Module (CM) mock-up in Building 5 at MSC, practicing procedures with the Apollo docking mechanism in preparation for the scheduled Apollo 11 lunar landing mission. Collins is at the CM's docking tunnel, which provides passageway to and from the Lunar Module (LM) following docking, and after removal of the tunnel hatches, docking probe and drogue. Credit: NASA.

"You'd try to figure out the main objective you'd want to accomplish with each one of these runs," Honeycutt said, "and then what other two or three things you'd want to throw in to make it exciting for a couple of other console positions."

They conducted two-day sims to give the flight controllers practice for handing things over to the next shift. They came up with sim problems for the entire building, where they would flip an electrical breaker and monitor how long it would take people to track down and solve the problem.

View of Mission Control. The Simulation Control Area can be seen through the glass windows on the far side of the room. Credit: NASA.

The Lunar Module Simulator in Building 5 at the Manned Spacecraft Center. Credit: NASA.

Honeycutt's favorite sim targeted Steve Bales, the Guidance Officer in Mission Control. Bales sat at the end of the front row—known as The Trench—and the night before an LM landing sim, one of the sim technicians would tie a string to the circuit breaker under the suspended flooring that was connected to Bales's console.

"Right in the middle of the sim, when Bales was coming up to make this really critical call, we yanked the circuit breaker and took all the power from his console," Honeycutt said. "Bales just told everyone to move down, and they all just shifted down one console to the left, like nothing had happened, and they did the landing. It was really something to witness, but it was an opportunity to get everybody involved."

At the end of each sim exercise, the SimSup would hold a debriefing with everyone to provide an overview of how well the flight control team performed or how effective they were in chasing down problems—but these debriefings were always conducted in the spirit of building the individuals into a cohesive team.

The Apollo 4 Command Module, with flotation collar, is hoisted aboard the USS Bennington, *recovery ship for this uncrewed, Earth-orbital space mission. The Command Module splashed down on November 9, 1967, 934 nautical miles northwest of Honolulu, Hawaii, in the mid-Pacific Ocean. Credit: NASA.*

The prime crew of the Apollo 8 mission training for recovery operations in the Gulf of Mexico. Left to right are astronauts William A. Anders, Lunar Module pilot, and Frank Borman, commander. A team of MSC swimmers assisted with the training exercise. Credit: NASA.

With missions scheduled approximately every two months, the two identical Mission Control Rooms were always busy running simulations for two different missions. In 1968, the sim team usually worked seven days a week, from 10 to 12 hours a day. Three days a week they'd run sims for Apollo 7 and the other three days they'd run them for Apollo 8.

"That left Sundays to figure out what we were going to do next week," said Honeycutt. "It was pretty hectic sometimes from a family-life point of view. But you couldn't have blown us out of there with a stick of dynamite. Everyone was having so much fun, and things were moving so fast. Nothing was the same from flight to flight, because every flight had a different set of objectives, and the envelope got expanded on each flight, so there was more to do. It was wonderful."

ALL OF NASA'S SPACECRAFT UP TO THIS
point had returned to Earth by splashing down in the ocean. Therefore, NASA could rely on that rich history of experience when it came to fishing the Apollo CM out of the water and picking up the crew. But Apollo was bigger and heavier than any previous spaceship, so NASA's Landing and Recovery Division had some challenges to deal with in order to get ready for the first crewed Apollo flights.

"In training for Apollo recovery operations, we didn't do simulations quite like the astronauts and flight controllers did," said Milt Heflin, a NASA landing and recovery engineer. "Instead, we had about four or five different mock-up versions of the CM, and one way we'd train was to put them in the Gulf of Mexico using an old flat-bottom Army landing craft. We became very adept at what we needed to do by testing it over and over again."

NASA worked with the Department of Defense to utilize aircraft carriers for the spacecraft recovery. Swimmers from the Navy Underwater Demolition Team (later called the Navy SEALs) would leap into the water from a helicopter to wrestle a flotation collar around the capsule, pop the spacecraft hatch and help the astronauts make it safely into the recovery helicopter. NASA's Landing and Recovery Division would work with Underwater Demolition Teams in San Diego and train at the naval base there, as well as on board aircraft carriers.

"We really had a real worldwide force through the Department of Defense to support us," Heflin said. "The Navy and Air Force would send representatives to the Manned Spacecraft Center, where we would meet prior to every mission, and go over the details of what was required to do the job. Then we would send our teams from the Landing and Recovery Division out to conduct briefings and training on board the ships and at aircraft staging bases that would be conducting the spacecraft recovery."

The Biological Isolation Garment (BIG) worn during a qualification test. Credit: NASA.

Milt Heflin atop the Apollo 8 command module prior to "safing" (removing hazardous propellants) at Ford Island hangar on Oahu on December 29, 1968. Image courtesy of C. Mac Jones.

Even though the CM weighed about 12,000 pounds, in the water it acted like a cork, bobbing and weaving amid the waves. The shape and size of this cork presented a large surface area to the wind, causing it to be blown downwind.

"When we would brief a new team of the Underwater Demolition Team frogmen," said Heflin, "we'd tell them, 'Now, we know you guys are really strong swimmers, but you cannot swim as fast as this Command Module is going to drift across the surface of the water, so don't try, don't even try.' Sometimes they learned the hard way to not jump out of the helicopter unless they were downwind of it, with the CM heading toward them. If they dropped a little upwind, they could never catch it."

Heflin said the great thing about using the big attack carriers (like the USS *Essex*, which would be used for Apollo 7, and the USS *Yorktown*, which would be used for Apollo 8) was that, as the carrier would come alongside the spacecraft, the giant ship would create a leeward, calm area in the ocean. This made the task of raising the CM out of the water with a specialized crane much easier.

"With this being a 12,000-pound object that you're trying to bring up out of the water, it has a tendency to start swinging like a pendulum," he said. "And of course, as the line gets shorter, the pendulum wants to swing faster. We would have sailors using tending lines that had been hooked up to the Command Module prior to being lifted out of the water so that as it came out of the water, the sailors could steady it. There was a trick to it, though, and that was probably where we had to train the most."

Part of Heflin's responsibilities for the upcoming missions that would land on the Moon included overseeing the development of the Biological Isolation Garment (BIG), a containment suit for the astronauts to prevent any contamination in the unlikely case they had been infected with some sort of lunar life-form. The BIGs would be tossed into the spacecraft after the hatch was opened following splashdown, and the astronauts would stay in the BIGs until they were sealed inside their Mobile Quarantine Facility, which was a specially outfitted Airstream trailer.

To choose the material for the BIGs, Heflin worked with the biological weapons testing facility at Fort Detrick in Frederick, Maryland, as they could ascertain if certain materials would contain a specific size of microorganisms. The experts there recommended a rubber-coated cotton material and designed a coverall-type garment with an attached breathing apparatus, which looked like a gas mask.

Once a prototype suit had been constructed, Heflin decided he needed some additional data on the BIGs, because he was worried about how hot the astronauts would get while wearing the suits and if they'd be able to communicate through the bulky gas masks.

"We had the go-ahead to get the work done and do whatever we needed to do," said Heflin. "I had an idea for a test, and one of my colleagues agreed to do it: He put on one of the BIGs, and we had a medical doctor insert a rectal probe, and then we sent this guy in the suit outside to sit in the Houston humidity and sunshine in one of those classic gray government chairs. The doctor was right there, about 5 feet away with a device that would measure this guy's core body temperature. I'm sure it was almost like torture because he indeed did get hot. But we got data we needed right away and it didn't cost anything."

The concern with communications came in particular when the astronauts were getting picked up by the helicopter, where there would be prop wash and noise. Heflin wanted to use a battery-operated microphone apparatus for crew communication, but of course, it needed to be tested first. One of Heflin's coworkers had a pickup truck with a roll bar, and Heflin convinced the truck owner to stand up in the back of the truck, hold on to the roll bar and stick his head above the cab while wearing the gas mask and microphone. They drove 60 miles an hour (97 km/h) down the street to simulate the prop wash and wind noise. Again, they got the data—and at very little cost.

"Quick and easy always did the trick," Heflin said. "But I do believe the BIG was the ugliest thing I've ever seen anybody ever have to wear. We just hoped it did the job because saving the world from lunar bugs was something nobody ever had to do before."

Warning system engineer Jerry Woodfill on the far left in the Mission Evaluation Room during an Apollo mission. Credit: NASA.

WHAT IF TWO THRUSTERS GOT STUCK

in the on position? What if the thrusters overheated due to heavy use during landing? Would this generate an alarm during the crucial landing or postlanding phase of the mission?

Jerry Woodfill spent another night lying awake, thinking about the Caution and Warning System onboard the Apollo spacecraft. It was his system—he was the lead engineer—and in his mind, the alarm system personified all the work that had been done since the Apollo 1 fire in regard to crew safety. But what-if questions about alarms for the LM thrusters kept nagging him.

The Caution and Warning System monitored critical conditions of the spacecraft, and any malfunction or out-of-tolerance condition—such as problems with the propulsion systems, environmental control or the guidance computer—caused a status light to illuminate. It also activated the master alarm, sending an alarm tone to the astronauts' headsets. The master alarm light and tone would continue until one of the crew reset the master alarm circuit.

"My system had two categories of alarms: one, a yellow light for caution when the astronaut could invoke a backup plan to avoid a problem, and the other, an amber warning indication of imminent life-threatening failure," Woodfill explained. He worked with engineers and managers on all the other parts of the spacecraft that the warning system monitored for failure to set the "trip level" of what conditions would set off an alarm. He also conferred with the astronauts on writing up the instructions for their in-flight operational checklist so they understood the alarms and knew what to do should they sound.

What Woodfill calls "nuisance alarms" were the most difficult to figure out. This type of alarm came on when no actual problem existed. For example, simply switching on a system might trigger an alarm until the system reached its operational state. Each instance of such an alarm had to be known by the crew and flight control team so that the occurrence of a nuisance alarm would not mask an actual problem. Worse, an errant alarm might confuse the crew or distract them during important moments of the flight.

Scores of such potential alarms became Woodfill's most challenging role as the warning system engineer for both the lunar lander and CM. Woodfill was scheduled to fly to New York to meet with Jim Riorden, the Grumman manager of instrumentation. The two men would go over the LM's Caution and Warning System, so Woodfill took the opportunity to discuss some of those middle-of-the-night contemplations.

"Jim, what would happen after the lunar landing when the heat from the hot engine warms our temperature sensor in the landing radar warning circuit?" Woodfill asked. "Might that trigger the landing radar alarm unnecessarily while the astronauts are out on their EVA? They'd have to return to the LM, cutting their exploration short to discover the alarm's cause. They'd find nothing more than a nuisance alarm. After landing no one cares about the landing radar. It had served its purpose."

"Jerry, you may have something there," Riorden said, pondering the idea. "We'll run a thermal analysis on the landing radar environment from touchdown to lift-off."

A week later, Riorden called Woodfill, saying that the voice in his head had been correct. That nuisance alarm would very likely occur under the postlanding conditions, as Woodfill surmised. The cost to correct the problem was negligible—one wire removed from the warning system.

Woodfill tried to imagine what NASA management and the news media would think if the first astronauts on the Moon were ordered to end their exploration hours early due to a nuisance master alarm. He ran some calculations on the estimated cost to the program, and he very likely earned his lifetime salary with NASA many times over in those few minutes of discussion with Riorden.

AS NASA PREPARED THE APOLLO 7 spacecraft for the first crewed flight in Earth orbit, the Soviet Union launched a series of robotic spacecraft called Zond that looped around the Moon and flew back to Earth. Then new intelligence photos from the CIA appeared to show the USSR preparing its own crewed lunar mission. These events prompted an August 1968 meeting between George Low, Robert Gilruth and Samuel Phillips (head of the Apollo Program Office at NASA Headquarters). Low suggested something bold: If the Apollo 7 mission was a success, change Apollo 8's flight plan from another Earth orbit mission to a circumlunar mission and have the crew orbit the Moon. At first, Phillips and Gilruth expressed shock at such an idea, but they quickly became intrigued. Delays with the LM meant it wouldn't be ready for Apollo 8's flight anyway, Low argued, and to send an Apollo flight around the Moon just might preempt any attempt by the Soviets to get to the Moon first. It would build momentum for NASA and advance scientific knowledge of the Moon. A future landing site could be observed and documented.

But would the flight hardware and the crew be ready for such an audacious mission?

More meetings ensued, adding the input of Christopher Kraft, Gene Kranz, Deke Slayton and other mission managers and planners, including Wernher von Braun, since this would involve sending the first crewed flight of the Saturn V to the Moon. Was all of this even remotely possible? Almost everyone who heard the idea had the same reaction: initial shock quickly turning to intrigue. This daring, gutsy move would be absolutely pivotal for meeting the "in this decade" goal, as well as getting to the Moon before the Russians. But it also—if it was successful—would very likely advance the timeline for a mission to land on the Moon. Deke Slayton called in Frank Borman, currently training as commander for Apollo 8. In a matter of minutes, Borman was all in. He and his crewmates Jim Lovell and Bill Anders met with Christopher Kraft, Bill Tindall and several members of the Mission Planning and Analysis Division team, and in one afternoon they outlined the basic parameters of the mission. A launch window for December 21 of that year offered the first chance that worked for the launch teams, recovery teams and all the other mission planners. Preparations went ahead, and all those who had been briefed about this new Apollo 8 mission were told to keep it secret. But, as Gene Kranz later noted, "it was like trying to hide an elephant in your garage."

View of activity in the Mission Operations Control Room in the Mission Control Center. The flight director's console is in the foreground. Eugene F. Kranz, chief of the MSC Flight Control Division, is in the right foreground. Seated at the console is Glynn Lunney, head of the Flight Director Office, Flight Control Division. Facing the camera is Gerald Griffin, flight director. Credit: NASA.

IN THE MEANTIME, THE REST OF MSC

was preparing for Apollo 7's important flight. Gerry Griffin would have his first job as flight director for this mission, along with Gene Kranz and Glynn Lunney. The three flight directors agreed they were looking forward to working with the crew of Wally Schirra, Donn Eisele and Walt Cunningham. However, during some of the first integrated simulations with the crew, Lunney noticed that Schirra directed some hostility toward the SimSup and that he seemed unwilling to go through several of the training scenarios. The tensions seemed to defuse after Lunney indicated to Schirra that this lack of cooperation might need to be reported to chief astronaut Deke Slayton. Lunney thought that was the end of any problems with the crew. But after the mission launched, Lunney realized he should have seen this as a warning sign.

On Apollo 7's launch day, October 11, 1968, the stakes were high. It was the first time NASA sent astronauts into space since the Apollo 1 fire twenty-one months prior, and this mission would be a crucial test of the new and improved CM. During its ten-day flight, Apollo 7 would need to prove all the hardware could function properly to enable a mission to the Moon.

Apollo 7 lifts off from Cape Kennedy Launch Complex 34 on October 11, 1968, for the first crewed lunar orbital mission. Credit: NASA/KSC.

Astronauts Wally Schirra (on right), mission commander, and Donn Eisele, Command Module pilot, are seen in the first live television transmission from space. Schirra is holding a sign which reads, "Keep those cards and letters coming in, folks." Out of view at left is astronaut Walter Cunningham, Lunar Module pilot. Credit: NASA.

The Apollo 7 crew after splashdown, arriving aboard the USS Essex. Left to right are astronauts Wally Schirra, Donn Eisele, Walt Cunningham and Dr. Donald E. Stullken, NASA Recovery Team leader from the Manned Spacecraft Center's (MSC) Landing and Recovery Division. Credit: NASA.

It was a hot day at Cape Canaveral when the Saturn IB lifted the crew to orbit in a perfect launch, a great new beginning for Apollo. The first day, the crew conducted tests on equipment and procedures, with only minor problems. But by the next day, complaints from the crew began about the food, the packed and demanding schedule and the cumbersome waste-management system for bathroom needs. Then, although all the systems were operating well, the crew came down with colds. They soon found that having a stuffed head in weightlessness can be problematic, since nasal passages don't drain well.

Of course, with public interest in this first crewed Apollo flight running high, the media grabbed the story about the head colds but also soon noticed abrupt, terse exchanges between Apollo 7 and the ground.

There was a request from Mission Control to power up the new TV camera system ahead of schedule to check the circuits. Schirra refused. "You've added two burns to this flight schedule, and you've added a urine water dump, and we have a new vehicle up here, and I can tell you at this point TV will be delayed without any further discussion until after the rendezvous."

However, later, when it was time for the first live television broadcast from a US spacecraft, viewers around the world were treated to the Walt, Wally and Donn Show, a lively, spirited and happy exchange from inside the CM. The smiling crew displayed hand-lettered cards reading, "From the Apollo Room, High Atop Everything" and, "Keep those cards and letters coming in, folks." The show was a hit—not to mention a public relations win for NASA—and all seemed well.

Meanwhile, the CSM performed superbly except for some minor problems with the fuel cells. But the griping continued from the crew, with complaints about noisy fans and water puddling from a coolant line, along with more rebuffed requests when ground controllers added or changed a task. In what should have been a triumphant first flight, frustrations ran high both in space and on the ground.

A final breach of protocol came at the end of the mission when the crew didn't want to wear their helmets for reentry due to their stuffy heads. Finally, in a rare move, Slayton himself radioed up to the astronauts, telling them all other crews donned their helmets for landing and there was no previous experience in landing with them off. After a testy exchange, Slayton finally said, "It's your neck, and I hope you don't break it." Schirra signed off abruptly with, "Thank you, babe."

The crew landed safely without their helmets. However, they never flew in space again.

Still, this was an essential mission for Apollo, even with a grumpy crew. The flight itself was nearly perfect, circling Earth 163 times. The mission met more objectives than originally planned, the spacecraft performed well and the *Walt, Wally and Donn Show* would win an Emmy Award. Now it was time to focus on Apollo 8's daring mission and continue to prepare for an eventual mission to land on the Moon.

NINETEEN SIXTY-EIGHT HAD BEEN A

turbulent year so far, for both the US and the rest of the world. Assassins killed Martin Luther King Jr. and Robert Kennedy. The Vietnam War continued to escalate, and details of the bloody Tet Offensive and My Lai Massacre were brought into American living rooms by Walter Cronkite and other reporters on the scene. Protests by antiwar demonstrators, students and counterculture activists resulted in violence at the Democratic National Convention in Chicago, on college campuses and in the streets. There were prison riots and rising drug use. "Don't trust anyone older than thirty," said the free-speech movement. Soviet tanks in Czechoslovakia prompted the US to cancel arms treaty negotiations with the USSR. Students protested in Paris. Hundreds of students were killed in Mexico.

Apollo 7's success provided a brief respite from the horrifying news reports. But then came Apollo 8.

On December 21, Apollo 8 took off atop a Saturn V rocket from Launchpad 39A at Kennedy Space Center. After one and a half orbits around the Earth, the third-stage booster began a four-minute burn to put the spacecraft on a lunar trajectory. Over the next few hours, Anders, Borman and Lovell experienced an unusual sensation that no human had ever encountered before: At first, they could see all the way from Chile to England; then the Earth became an entire sphere; then they watched in wonder as the whole Earth began to get smaller and smaller.

The crew of the Apollo 8 mission, left to right, are James A. Lovell Jr., Command Module pilot; William A. Anders, Lunar Module pilot, and Frank Borman, commander. They are standing beside the Apollo Mission Simulator at the Kennedy Space Center. Credit: NASA.

Apollo 8 coasted to the Moon. During a 2:00 a.m. shift in the Flight Dynamics staff room, Jack Garman and his fellow engineers took bets. In all the Apollo missions so far, the Guidance and Navigation Computer worked on a coordinate system based on the Earth, performing calculations for navigation, gravity and position, all based on Earth's gravitational influence. But this flight was the first time a spacecraft with humans on board would go far enough away from our world to enter the gravitational sphere of influence of another celestial body: the Moon. Garman wanted to know the exact moment that switch took place. Nothing dramatic would happen, no jolt or noticeable change to the crew, but a shift for the computer and a first for humanity as the Moon's gravitational force on Apollo 8 became stronger than the Earth's.

A striking view from the Apollo 8 spacecraft showing nearly the entire Western Hemisphere, from the mouth of the St. Lawrence River, including nearby Newfoundland, extending to Tierra del Fuego at the southern tip of South America. Central America is clearly outlined. Nearly all of South America is covered by clouds, except the high Andes Mountain chain along the west coast. A small portion of the bulge of West Africa shows along the sunset terminator. Credit: NASA.

Since Apollo 8 was a mission full of firsts like this, the flight dynamics team set their consoles to notify them when certain events took place—such as when the translunar injection burn successfully put the spacecraft on course for the Moon and other milestones. An event light on their console would quietly let them know when the shift in gravitational influence happened.

"Now, navigationally, we knew exactly the point the vehicle was actually going to cross this threshold," Garman said, "but we're down to a gnat's hair here, because there's close to a two-second lag time in telemetry due to how far away the spacecraft was. We're trying to take bets on exactly when that light is going to come on."

Fifty-five hours and forty minutes into the flight, when Apollo 8 was 38,759 miles away from the Moon, the light came on. The engineers just stared at it.

"My God," Garman told his colleagues. "My heavens, it's real. They're falling toward the Moon."

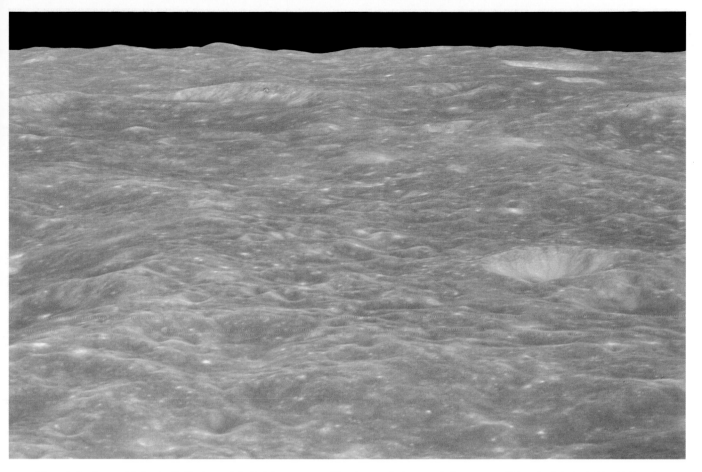

After entering lunar orbit, the Apollo 8 astronauts looked down on rugged terrain never before seen by human eyes. This scene is typical of farside terrain, consisting of craters superimposed on older craters. The view extends about 350 statute miles to the horizon. Credit: NASA.

JOHN PAINTER RETURNED TO THE HOUSTON area for the Christmas holiday and got in touch with his old colleague Howard Kyle. Kyle invited Painter to join him in the Vehicle Systems Staff Support Room in Mission Control during a critical mission event, the lunar orbit insertion burn. If the engines fired successfully, Apollo 8 would become the first crewed spacecraft in lunar orbit. Kyle gave Painter the opportunity to sit at a console and listen on a headset to the Apollo USB voice channel, the system he had worked so hard to document and refine.

The chatter on the communication loop indicated Apollo 8's precise trajectory was about to send them around the far side of the Moon. If all went well, the SM's service propulsion system engine would fire just long enough to slow their velocity, allowing the Moon's gravitational field to capture the spacecraft and put them in lunar orbit. With the crew behind the Moon, no communications were possible during the engine burn. If the maneuver was successful, Mission Control would regain the signal after thirty-two minutes and thirty-seven seconds. If the engine didn't fire at all, they would regain the signal in twenty-two minutes; it would also mean Apollo 8 was heading back to Earth. But a variety of engine malfunctions and firing times could result in different signal reacquisition times.

The Apollo 8 photo that has come to be known as "Earthrise," the first look by human eyes of our home planet from another world. Credit: NASA.

DURING THE FIRST THREE ORBITS, APOLLO

8's astronauts kept the spacecraft's windows pointing down toward the Moon while they hurriedly filmed and photographed the craters and mountains below. One of their main tasks was reconnaissance for the future Apollo landings.

On the fourth swing around from the Moon's backside, Borman rolled the spacecraft to a different orientation, pointing the windows toward the horizon to get a navigational fix. A few minutes later, he spotted a blue-and-white object coming over the horizon.

"Oh, my God! Look at the picture over there. Here's the Earth coming up," Borman shouted. This was followed by a flurry of exclamations by Anders and Lovell and a scramble to find a camera and color film. Borman took the first image in black-and-white, showing Earth just peeping over the horizon. Then Anders found a roll of 70-millimeter color film for the Hasselblad camera.

This was the first look by human eyes of our home planet from another world, a beautiful, heart-catching sight that sent a torrent of sheer wonder surging through the crew. The Earth was the only thing they could see that had any color to it: Space was black, the Moon was gray. Lovell called Earth "a grand oasis in the vast loneliness of space."

Painter listened as the voice channel became silent as the spacecraft passed behind the Moon, the headset producing the slight hiss of an empty channel. Everyone watched the clock. Twenty-two minutes passed with no signal. After earlier chatter, the voice channel again became silent as everyone waited for the time of reappearance.

More silence. Then, a familiar sound came over the radio channel. It was a descending tone, ending in a unique sound that Painter had heard many times before: *wheeeeeeouuu . . . thunk.* It was the sound of the ground's USB radio receiver correcting for the Doppler shift of the incoming radio signal and then locking onto that signal. Then came a voice, Lovell's simple call, "Houston, Apollo 8. Burn complete." From Mission Control, Capcom Jerry Carr replied, "Apollo 8, this is Houston." The room broke into cheers, and Painter's eyes got a little wet.

AT THE MOON, THE APOLLO 8 CREW

prepared for a special live thirty-minute broadcast. On Earth, it was Christmas Eve. Millions of people tuned in, watching and listening with rapt attention, seeing the grainy footage of the Moon, hearing the astronauts describe their lunar views. Then the portable television camera showed the blue marble of Earth, hanging in the blackness of space.

Bill Anders said, "We are now approaching lunar sunrise, and for all the people back on Earth, the crew of Apollo 8 has a message that we would like to send to you. 'In the beginning, God created the heaven and the Earth.'"

The crew took turns reading the first ten verses of the book of Genesis in the Bible, and Borman ended with, "And from the crew of Apollo 8, we close with good night, good luck, a Merry Christmas—and God bless all of you, all of you on the good Earth."

The Mission Control Room on the third day of the Apollo 8 lunar orbit mission. Seen on the television monitor is a picture of Earth, which was telecast from the Apollo 8 spacecraft 176,000 miles away. Credit: NASA.

At that moment in the Mission Control Center, throughout all the backroom support teams, in the packed viewing room overlooking Mission Control where MSC family and friends had gathered, across the country at other NASA centers and contractor sites or wherever anyone was who had a hand in Apollo 8, more eyes became wet. "Hearing those words along with what we were seeing, and on Christmas Eve, it was just really, really emotional," said Norman Chaffee. "It was just magical."

On Christmas morning, Mission Control waited anxiously for word that Apollo 8's engine burn to leave lunar orbit had worked. They soon got confirmation when Lovell radioed, "Roger, please be informed there is a Santa Claus."

After the crew splashed down in the Pacific on December 27, they received hearty congratulations from around the world. Two telegrams stood out to Frank Borman. One came from several Russian cosmonauts saying, "We congratulate the American scientists and all the American people. We wish you further success on all other flights. We are confident future exploration of outer space will greatly benefit earthly men. We congratulate you on a successful step on a noble goal."

The second telegram, sent anonymously, simply said, "Congratulations to the crew of Apollo 8. You saved 1968."

CHAPTER 8

1969

I don't give a damn what you think, give me your data!

—DON ARABIAN, head of the Apollo mission evaluation room

NASA BEGAN 1969 WITH A RENEWED sense of optimism and determination. But even with Apollo 8's incredible success, there was no time to enjoy the moment. If anything, the Apollo program needed to pick up even more speed—and quickly.

"Everyone came out of Apollo 8 ready to go to work, and work hard," said Frank Hughes. "We felt like everything was coming together."

The little Moonship, the Lunar Module (LM), was finally ready to fly. Apollo 8 had proven the rest of the systems were Moonworthy, and Apollo 9 and Apollo 10 were quickly coming down the pike, each with their own set of objectives to make a Moon landing possible. Apollo 11 was now considered the earliest possible mission to attempt a lunar landing, but NASA repeatedly issued a caveat: Any problems arising on Apollo 9 or Apollo 10 could demonstrate the need to postpone the lunar landing to a later flight.

What Apollo 11 needed now was an official crew.

These three astronauts were selected by NASA as the prime crew of the Apollo 11 lunar landing mission. Left to right are astronauts Edwin "Buzz" E. Aldrin Jr., Lunar Module pilot; Neil A. Armstrong, commander; and Michael Collins, Command Module pilot. They were photographed in front of a Lunar Module (LM) mock-up beside Building 1 at MSC following a press conference. Credit: NASA.

Left: Apollo 11 Commander Neil Armstrong performing Lunar Module simulations at Kennedy Space Center. Credit: NASA.

Two members of the Apollo 9 crew participate in simulation training in the Apollo Lunar Module Mission Simulator (LMMS) at the Kennedy Space Center. On the left is astronaut James A. McDivitt, commander, and on the right is astronaut Russell L. Schweickart, Lunar Module pilot. Credit: NASA.

While the astronauts for Apollo 9 and Apollo 10 had been set for quite some time, NASA was waiting for the right moment to make an official announcement on Apollo 11's crew. Media speculation became rampant. Of course, it seemed obvious that Neil Armstrong and Buzz Aldrin would be named, as they had been the backup crew on Apollo 8, and in the foggy math of chief astronaut Deke Slayton's crew assignments, being backup usually meant you'd skip two missions, then be prime crew on the third. But the third spot on Apollo 11's crew was in flux, as Jim Lovell had been moved to Apollo 8 after Mike Collins needed surgery for a bone spur on his spinal column in 1968. Additionally, NASA had been shuffling crew assignments in all the missions ever since the deaths of See and Bassett.

Only after Frank Borman, Jim Lovell and Bill Anders triumphantly returned to Earth, were named *Time* magazine's Men of the Year for 1968, spoke before a joint session of Congress and were greeted at the White House by outgoing President Johnson—all in the span of less than two weeks—did NASA schedule a press conference to make the announcement for Apollo 11's crew. On January 10, the

world came to know Apollo 11's astronauts as commander Neil Armstrong, Lunar Module (LM) pilot Buzz Aldrin and Command Module (CM) pilot Mike Collins. Media questions quickly shifted to who would be the first to step on the lunar surface, and Slayton, sitting up on the dais with the astronauts, told reporters such a decision had not yet been made.

The first launch window for Apollo 11's flight came in July, and although Armstrong, Aldrin and Collins had been training for Apollo for quite some time, they quickly began their specific training for Apollo 11's mission to land on the Moon. The crews of Apollo 9 and Apollo 10 had already been taking their turns in the simulators, repeatedly and methodically practicing every maneuver and key punch required for success. Apollo 9 was considered a test flight, but it would be an engineering test flight like no other; one could argue it was the most significant engineering flight of the Apollo program. The ambitious mission conducted in Earth orbit entailed many Apollo-era firsts, including the first crewed flight of the LM and the first rendezvous of the Apollo program, as well as checkout and testing of all the systems on the Command and Service Modules (CSM) and almost all the systems of the LM.

Apollo 9's Jim McDivitt, Dave Scott and Rusty Schweickart were the first of the Apollo crews to use the two mission simulators for both the CM and LM. By the time of Apollo 9's launch, the crew had spent more than seven hours in training for every hour of their scheduled ten-day (241-hour) mission. Still, they felt like they couldn't get enough training, especially for the rendezvous. Schweickart also trained for the first in-space test of the spacesuit for spacewalks, the Extravehicular Mobility Unit, during his scheduled two-hour extravehicular activity (EVA) to simulate external transfer and rescue techniques.

Apollo 9 became the first mission since 1965 where astronauts hung nicknames on their two spacecraft. But it wasn't just for fun; it was a matter of logistics and safety. NASA hierarchy had raised their brows and issued an edict forbidding spacecraft nicknames after Gus Grissom christened his Gemini 3 spacecraft *Molly Brown* (as in *The Unsinkable Molly Brown*), referring to the sinking of his Mercury spacecraft.

"Well, here we are coming along with Apollo 9," said Rusty Schweickart, "now we've got two spacecraft, so when we're separated, what are we? Talking to each other, are we Apollo 9 Alpha and Apollo 9 Bravo, or whatever? And then when I go outside on EVA, I'm sort of a third spacecraft because now we're communicating over the radio with three different things. So, am I just Rusty or what am I?"

Apollo 9 prime and backup crews practice simulated altitude chamber run and egress with the CSM and LM at the chamber "L" at KSC. The Command Module is still covered with the protective blue coating. Credit: NASA.

To avoid confusion, McDivitt, Scott and Schweickart decided they needed to have calls signs with no ambiguity, something clear and distinct. Their initial plan was to come up with names that weren't humorous; they had to be something very obvious and bland, names that no one at NASA could complain about. Nonetheless, they ended up making their nickname decisions after dinner and a few drinks one night in Downey.

"You'd look at the Command Module on the factory floor and it had a thin blue coating on it, and it looked like a gumdrop," said Schweickart. "Well, how can anyone complain about gumdrops? And when you look at the Lunar Module, what else does it look like but a spider? And when I would step outside the spacecraft and do the EVA, the logical name was *Red Rover* [Schweickart has red hair]. So, *Gumdrop, Spider* and *Red Rover*. We didn't ask anybody, we didn't tell anybody, we just started using it in the simulations. As we got closer and closer to flight, Mission Control started using it and it stuck."

From then on, the Apollo spacecraft nicknames and call signs came back to stay.

Apollo 9 launches from Pad A, Launch Complex 39, Kennedy Space Center at 11:00 a.m. (EST), March 3, 1969. Credit: NASA.

On Apollo 9 launch day, March 3, 1969, the crew's main emotion was "at last." They had been chosen for this flight more than two years prior, but the delay after the Apollo 1 fire and a short delay to make sure the Apollo 9 crew didn't have colds meant they were more than ready.

"You go through so many simulations and so much training, and you've sat in that spacecraft so many times," Schweickart said. "It seemed like it was never going to happen."

The launch at precisely 11:00 a.m. from Cape Canaveral was perfect. The crew felt the now-expected vibration and power from the Saturn V rocket, rising slowly and majestically then gaining speed and noise. It was a flawless, perfectly normal launch. Except for one thing.

North American Rockwell artist's concept illustrating the docking of the Lunar Module ascent stage with the Command and Service Modules during the Apollo 9 mission. Credit: NASA/North American Rockwell.

When the crew had entered the spacecraft, lead launchpad technician Guenter Wendt had tightened their shoulder harnesses so securely they couldn't move at all. During the long countdown, their arms became nearly numb, so Scott and Schweickart loosened their harnesses, just a tad, to afford their arms some freedom of motion. But at two minutes and forty-two seconds into the flight, engine cutoff of the first stage of the Saturn V rocket occurred. The spacecraft went from almost 7 million pounds of thrust to zero in about a millisecond. The sudden lack of acceleration threw the crew forward. McDivitt's body jerked just slightly forward.

"Dave and I, well, we flew forward and our helmets stopped probably an inch from the instrument panel," Schweickart said. "We looked at each other and said, 'Whew. We'd better tell the next guys not to loosen anything.' Guenter knew what he was doing."

The first two days of Apollo 9 went smoothly. The astronauts docked the two spacecraft successfully on the first try and later fired *Gumdrop*'s engines several times with *Spider* attached, proving they could easily maneuver the two spacecraft in docked formation during orbital maneuvers.

Day three had McDivitt and Schweickart donning spacesuits so they could enter the LM and turn on all the systems, but as soon as Schweickart finished getting into his suit, he threw up.

Schweickart had worried this was going to happen. He had gotten sick during most of the Vomit Comet flights, and tried to take things easy the first couple of days in space because of rumors of what had happened to Frank Borman during Apollo 8.

Space motion sickness, or space adaptation syndrome, as it would later be named, was unknown to the US space program until Apollo 8. The nausea, headaches and occasional disorientation that about half of all space travelers experience comes from the body—in particular the inner ear—adapting to the change in gravitational forces. The Gemini and Mercury spacecraft were small enough that no one could move around much—the astronauts joked they "wore" the spacecraft. On Apollo 7, no one reported any queasiness as a result of being in space, but on Apollo 8, Borman had gotten sick. He tried to hide it from Mission Control and wouldn't talk about it after the flight.

"I'm not sure how he thought he was going to hide the barf bags in the spacecraft," Schweickart said, "but I guess getting sick in space isn't part of having 'the right stuff.' For all kinds of reasons, which are Frank's, he wouldn't really come forward with it. He wouldn't do any tests afterward, so therefore we didn't know anything about it."

The price of not knowing anything about space motion sickness meant Schweickart got sick on the very day he needed to function the most. It turns out, as doctors determined later, the best way to prevent space motion sickness is for an astronaut to move their head in a relatively controlled manner so they come up short of being sick; the controlled motions give the inner ear a chance to adjust to the way fluids move within the body in microgravity. This accelerates the adaptation, which usually passes after the first few days of a spaceflight.

Schweickart had spent most of the first two days in space strapped to his seat, going over checklists, not moving much in order to keep from getting sick. But getting into a spacesuit in space takes gymnastic-like gyrations, which prompted his nausea. Schweickart barfed again later while working the LM with McDivitt.

"Looking to the next day and the idea of doing an EVA outside the spacecraft was daunting," Schweickart said. "Getting into a spacesuit and helmet, being in zero-g and then barfing are totally incompatible. I don't want to be too graphic about it, but you can't get your hand up to your mouth or nose to clear stuff away and it is very sticky stuff. In essence, if you are in a spacesuit and you barf, you could end up dead, suffocated. At the very least, it would be a very ugly scene. Not knowing whether I would be sick or not, we decided to cancel the EVA."

Thinking ahead, Schweickart knew if he continued to be sick, they might have to cancel the rest of the mission. That meant they might not be able to test the rendezvous and docking maneuvers and the rest of the flights down the line would be changed, possibly delayed. Schweickart went to sleep that night thinking that he personally could be the reason that NASA didn't meet Kennedy's goal of landing on the Moon before the end of the decade.

For Schweickart, it was a low point, in not only the mission but in his entire life.

On day four, after conferring with Houston, the astronauts decided to go into the LM. The plan was to execute the entire checklist before and after the EVA but to not depressurize and do the EVA itself. But while they were in Spider, McDivitt kept an eye on Schweickart and noticed he was obviously feeling better, working inside his spacesuit with no problems. McDivitt looked over at Schweickart.

"You know, you're looking a lot better today. How are you feeling?" McDivitt asked.

"I'm feeling a lot better," Schweickart replied.

They looked at each other, each hesitating a moment, and McDivitt finally said, "Well, let's just keep going and we'll see what happens."

About ten minutes from the time of depressurization, McDivitt said, "What do you think?"

The two knew each other well enough that Schweickart could say, "I think we are a go."

McDivitt hit the transmit button and said, "Houston, Apollo 9. We are going to go ahead with the EVA."

Schweickart performed a shortened version of the original spacewalk plan. He would go from his lowest moment to the peak experience of his life in less than twelve hours.

Rusty Schweickart during his extravehicular activity on the fourth day of the Apollo 9 mission. The Command/Service Module and the Lunar Module 3 Spider are docked. This view was taken from the Command Module Gumdrop. Schweickart is holding a 70 millimeter Hasselblad camera. Credit: NASA.

"So then we do the EVA and I get this opportunity to go outside the spacecraft, which is really a special thing," said Schweickart. "And then, in a totally unrelated set of circumstances, Dave Scott's movie camera jams, and I have five minutes to just take some time to appreciate where I was and consider what all this meant. I had the chance to be out there, not as a NASA astronaut but as a human being. For five minutes I got to look at the Earth below and the black space above and the sun over my shoulder. It was an incredible, spectacular view."

Schweickart told himself to just absorb the moment, soak this up. Later, he would come to think of those five minutes in philosophical, existential terms that would profoundly change his life. But now, with Scott's camera working again, it was time for Schweickart to get back to work and finish his EVA tasks, retrieving thermal samples from the exterior of the spacecraft and testing the spacesuit.

It had turned into a good day in space, and NASA's public affairs officer Jack Riley commented during NASA's broadcast of the events, "You heard it here, live, firsthand—the adventures of Red Rover and his friends, Spider and Gumdrop."

The Apollo 9 spacecraft approaches splashdown in the Atlantic recovery area to conclude a successful ten-day Earth-orbital space mission. Splashdown occurred at 12:00 p.m. (EST), March 13, 1969. Credit: NASA.

Day five brought the most-anticipated part of the mission: undocking the two spacecraft and flying *Spider* on its independent maiden voyage. McDivitt and Schweickart flew *Spider* for nearly six hours, rehearsing all the possible orbital phases of a real lunar mission. *Spider*'s engines fired flawlessly and smoothly.

"It actually handled kind of like a sports car," said Schweickart. "It was very crisp when you did rotational maneuvers, and it had a very snappy kind of response, which you want when you're going to land on the Moon, as you might have to make some last-minute maneuvers. So, *Spider* was a very fun vehicle to fly."

As Schweickart and McDivitt maneuvered the LM approximately 10 miles (16 km) below and 80 miles (129 km) behind the CM, Scott, still inside *Gumdrop*, watched in wonder as *Spider* became a small point of light in the distance.

Left: A view of the Apollo 9 Lunar Module photographed from the Command and Service Modules on the fifth day of the mission. Credit: NASA.

Spider returned for a rendezvous and docking to practice what would take place in the upcoming missions in lunar orbit, and the two spacecraft came back together cleanly and solidly. As everyone in Mission Control and the Mission Evaluation Room (MER) monitored the events in space, the performance of the spacecraft left no doubt that the LM was ready to make the lunar trip. "Apollo 9's uniqueness was in the technical and engineering aspects of making the Apollo Moon landings possible," said Schweickart. "We did our jobs, we showed the design worked well, and nothing really went wrong during our flight, in terms of spacecraft design."

The crew remained in orbit until March 13 and splashed down in the Atlantic Ocean north of Puerto Rico. They came down about 3 miles (5 km) from their recovery ship, the USS *Guadalcanal*, in full view of Milt Heflin and his landing and recovery team, who again worked their magic to bring the crew and the spacecraft on board.

The Apollo 9 crew on board the USS Guadalcanal *as they step from a helicopter to receive a red-carpet welcome. Left to right, are astronauts Russell L. Schweickart, David R. Scott and James A. McDivitt. Credit: NASA.*

Even with such a successful mission, questions remained about Schweickart's illness. Was his experience typical? What was the best way to adapt to being in space? Did NASA understand enough about this? Schweickart knew these questions had to be answered not only so that NASA could get a handle on how space motion sickness might affect upcoming missions but also so he could understand what he'd just been through. Schweickart made the tough decision to step out of line in the crew rotation schedule—likely giving up the chance for a later mission to the Moon—to undergo testing. His conscience told him it was the right thing to do.

"I didn't think we should be going to the Moon without knowing the answers," he said. "It wasn't clear to me if this was going to risk people's lives, because what if someone got sick while they were on the Moon? The only way we were going to be able to make a rational decision about committing with confidence to the upcoming lunar missions was to have more data."

He volunteered to be tested and shortly after Apollo 9 returned to Earth, Schweickart started spending weeks at a time at the Naval Aerospace Medical Institute in Pensacola, Florida.

WITH THREE MISSIONS NOW UNDER THE Apollo program's belt, everyone felt they were hitting their stride, especially the flight control team. This group of engineers was perhaps the most visible aspect of the Mission Control team; however, behind the scenes, several support teams provided real-time technical expertise for the controllers. Several different Staff Support Rooms (SSRs) were scattered around the Manned Spacecraft Center (MSC), and each was home to a large number of experts across all the different spacecraft disciplines.

One unique support room was the MER (the Mission Evaluation Room). Most other support teams were made up of experts for specific systems, but the MER housed a diverse team of engineering experts across several disciplines for both the spacecraft and the crew, primarily from the Structures, Propulsion and Power, and Guidance and Control Divisions. George Low formally organized the MER after the Apollo 1 fire, and it operated for all the Apollo flights. The technical-systems engineers assigned to the MER provided engineering expertise to whoever needed it, and the MER helped coordinate the efforts between MSC's Engineering and Development Directorate and engineering representatives from the spacecraft and system contractors who would be on-site while a mission was in progress. This included more than thirty subsystem managers and their contractor counterparts.

The Mission Evaluation Room in Building 45 during the Apollo 11 mission. Gary Johnson is wearing the striped shirt at the far end of the front table. Credit: NASA, image courtesy of Gary Johnson.

Mission Evaluation Room during Apollo 11. Credit: NASA.

The MER provided around-the-clock support during flights, using three shifts of personnel organized under individual shift team leaders. Those who manned the MER were all engineering experts for their systems; for example, Norman Chaffee regularly took shifts for propulsion, Gary Johnson was there to oversee the electrical systems and Jerry Woodfill monitored the Caution and Warning System.

While not in the same building as Mission Control (which was in Building 30), the MER was only about 300 yards (274 m) away and comprised most of the third floor of Building 45. During missions, the MER buzzed with activity and noise. One engineer likened its bustle to the commotion in a police precinct in a big city on a Saturday night. Even during nominal parts of the mission, or what others might consider downtime (such as during a crew sleep period), MER engineers constantly analyzed the incoming data from the spacecraft and shared any pertinent data with other support rooms and the flight control team. But during important points in the mission, such as orbital insertion maneuvers or lunar landing, everyone in the MER swung into action. Compared to the calm, ordered, military-like correctness in Mission Control, the MER, as one MSC engineer said, saw times when it seemed like everyone's hair was on fire.

Unlike the roomy, long rows of consoles in Mission Control, engineers in the MER sat shoulder-to-shoulder on gun-metal gray straight-backed steel chairs, positioned along both sides of large government-issue gray tables. Woodfill compared the arrangement to sitting "picnic-like around church basement potluck dinner tables." The MER needed to fit as many as one

hundred engineering experts at once. Sometimes, during key points of a mission, even extra engineers were on hand to monitor and observe critical events.

None of the engineers in the MER had individual console displays like the people in Mission Control; instead, data were displayed on 19-inch (48-cm) cathode ray tube (CRT) monitors that hung around the room near the ceiling. So that everyone could see the monitors, the MER was slightly darkened and heavy curtains hung over the windows. Still, it might be difficult to see the small numbers and letters on the monitors, so many MER engineers—especially those sitting on the far ends of the tables—used binoculars or spy glasses to better see the details on the screens. As far as keeping records of the ever-changing data displayed on the monitors, the engineers could acquire telemetry printouts from the Real Time Computer Complex (RTCC) data team, but that took time. The preferred method was to take a Polaroid photo of the screen. For the early Apollo missions, the Polaroid film would need to be pulled out and rubbed with a special emulsion. But by Apollo 11, MSC acquired the near-instant "picture in a minute" Polaroid cameras, which allowed the plastic covering over the photo to be pulled off after one minute. Because of the room's darkness and the small details on the photos, many in the MER used a specially designed magnifying glass with a flashlight attached to view the photos.

Most of the engineers, like Lonnie Jenkins who worked in the Propulsion and Power Division, meticulously hand-plotted the data on graph paper from each of the Polaroid snapshots and over the course of a mission would acquire a box full of pictures and a three-ring binder full of data plots.

Don Arabian (yelling into the phone) at the center of his table in the Mission Evaluation Room during the Apollo 11 mission. Credit: NASA.

Lonnie Jenkins's specially designed magnifying glass with an attached flashlight to view small Polaroid pictures of data captured from the monitors in the MER. Credit: Nancy Atkinson.

Having the data at their fingertips was crucial. "If we had an issue come up during a flight," said Jenkins, "one of the first things we did was establish a timeline. All the data had a time stamp and some of it was available in increments of microseconds. If we knew exactly when the problem started showing up and compared it with other data, we could see the whole picture of what happened and when it happened. We eliminated a lot of possibilities because of having an accurate timeline of our data and were able to enact fixes for any hardware problems."

The only "decorative" picture adorning the MER wall confirmed these engineers' devotion to data. It was a framed 8-by-11 piece of paper on which someone printed, "In God we trust, all others bring data." That was the MER motto.

But full-scale, hand-drafted technical drawings and blueprints were taped to the walls around the room—sometimes even out in the hallway—so that engineers could consult them whenever they needed.

MER staff wore headsets to listen to all the loops. But they could only listen, not transmit. A shared phone was available, two phones per table. Accompanying the headsets was a push-button audio channel selector box. Jenkins became particularly adept at being able to listen to all eight voice loops at once and could keep straight who was saying what. Somehow, he could ascertain how one conversation might relate to another or tune certain loops out.

"Lonnie was our data guru," said Chaffee, who worked closely with Jenkins. "He made sure everyone had the data that they needed, and he also coordinated all the troubleshooting for any anomaly. He saved the bacon for us many times by being able to give us the exact data we needed."

Any "hair on fire" activity in the MER was likely fueled by the man in charge, Don Arabian, chief of the Test Division for Apollo. Some called him Mad Don.

Arabian's main mode of communication was yelling. He would shout, swear, holler; whatever it took to get an immediate answer to his questions. His mind operated so quickly, he couldn't get the words out fast enough, and the faster he talked the louder he got. But everyone knew to listen because Arabian was always right.

Woodfill said Arabian commanded the MER team from his "throne-like" center seat of one of the long tables. "His loud, challenging voice could carry the entire length of the MER," he said. "Despite his fierce personality, he was a brilliant engineer. No one had a greater ability of assessing a spacecraft anomaly than Don."

Arabian would have a list of all the spacecraft anomalies and would check them off with gusto as the problems were solved. Whenever a problem came up, he wanted to know all the options, and he worked in absolutes. If anyone ever said, "I think this might be the problem," Arabian would retort, "I don't give a damn what you think, give me your data!"

Jenkins, quiet and reserved, was one of two people that Arabian never yelled at, but there were several people that seemed to be constant targets for Arabian's ass chewings. But Jenkins knew Arabian's temper was never meant as a personal attack. Jenkins compared Arabian's actions to the cartoon of a wolf and a dog that were constantly fighting, but when the noon whistle blew, they would sit down and peaceably eat lunch together.

"Don was like that," Jenkins said. "He could be chewing you out, but when lunchtime came, he would offer you a sandwich or an apple. Yelling and arguing was just his style. But he also respected your conclusions, as long as you had the data to back it up."

Outside of the MER, Arabian continued his high-energy ways, running a tight ship. His office was located on the third floor of one of the MSC office buildings, and when he wanted to talk to someone, he didn't bother picking up the phone—he'd just holler out their name.

"The guy he wanted most was one of his assistants, a guy named Bob Fricke," said Chaffee, whose office was on the same floor. "Whenever Arabian wanted Bob, he'd just holler out, '*Frickeeeeee!*'" The sound would resonate throughout the building, and Fricke would come running down the hall.

Even after a mission concluded, the MER remained important, as one of its primary assignments was to certify that all systems were ready and that it was safe to proceed for the next flight. After each flight, analyst managers like Jenkins met with managers for all the various systems and they would go over the mission report line by line, meticulously inspecting all the data gathered by the MER teams.

Throughout the Apollo program, the efforts of those in the MER resolved countless problems, constraints and crucial issues, ensuring mission success.

The crew of the Apollo 10 mission. Left to right are Eugene A. Cernan, Lunar Module pilot; John W. Young, Command Module pilot; and Thomas P. Stafford, commander. In the background is the Apollo 10 space vehicle on Pad B, Launch Complex 39, Kennedy Space Center, Florida. Credit: NASA.

IF FRANK HUGHES HAD TO CHOOSE A favorite crew, it would be the crew of Apollo 10: Gene Cernan, Tom Stafford and John Young. Hughes had spent a lot of time with this crew and got to know them well. The night before the mission launched, since there was no quarantine requirement before this flight, Hughes was able to have a drink with the crew. Suddenly Cernan said, "Okay, one more and then I need to hit the sack. Gotta go to the Moon tomorrow."

"Working with these guys was a delight," Hughes said. "They just had such a good time, they knew they couldn't land on the Moon, so there was no pressure. They always had a good time together getting ready for their dress-rehearsal flight."

An Apollo 10 view of crater Schmidt which is located at the western edge of the Sea of Tranquility. Schmidt has a diameter of 7 statute miles. The shadowed area is on the east side of the crater. Credit: NASA.

Replicas of Snoopy and Charlie Brown, the two characters from Charles Schulz's syndicated comic strip, Peanuts, *decorate the top of a console in the Mission Operations Control Room in the Mission Control Center, Building 30, on the first day of the Apollo 10 lunar orbit mission. Credit: NASA.*

Dress rehearsal. Test flight. Dry run. Practice mission. In NASA's world of sims, mock-ups and all-up testing, Apollo 10 became the most significant simulation run of them all. Besides giving the spacecraft and NASA's entire mission support team a near-complete workout, this mission provided a full checkout of the guidance and navigation system. It also gave the Manned Space Flight Network—the set of tracking stations around the world, built to support Apollo missions—a complete test in tracking and communicating with spacecraft at lunar distances.

On May 18, 1969, the Saturn V rocket sent Cernan, Stafford and Young on a near-perfect trajectory to the Moon in the spacecraft named *Charlie Brown* (the CSM) and *Snoopy* (the LM). Apollo 10 duplicated absolutely everything for a Moon landing—except for the landing part. Cernan and Stafford were to rehearse all the landing mission steps, up to the point where the LM would begin its powered descent. This would permit an all-up test of both the descent and ascent engines and other systems.

After a three-day cruise, Apollo 10 entered lunar orbit. A day later, Cernan and Stafford took *Snoopy* for a six-and-a-half-hour pathfinding test flight, going down to within 10 miles (16 km) of the craters and crags of the lunar surface.

"I'm telling you, we are low," Cernan excitedly radioed back to Earth. "We're close, baby! We is down among 'em!"

They flight-tested the LM's communications, propulsion, attitude-control and radar systems. They took numerous photos of the lunar terrain, especially the planned landing sites. They gathered data on the lunar gravity field to answer questions about the gravitational irregularities detected by the Lunar Orbiter and experienced by Apollo 8.

The crew reported that all systems were working perfectly. But when it came time to return to orbit to rendezvous with *Charlie Brown*, *Snoopy* suddenly tumbled out of control. On NASA's live broadcast feed, Cernan shouted, "Son of a bitch, what the hell happened?"

While flight controllers and the MER team back on Earth scrambled to look at their data to understand what was happening, Cernan and Stafford quickly realized one of them had put the rendezvous radar switch in the Auto position too soon. Instead of concentrating on how far above the Moon the LM was ascending, *Snoopy*'s radar began searching for *Charlie Brown*, causing the guidance system to flip the LM around wildly. Quickly, the two astronauts turned the switch off and jettisoned the descent stage to regain control of the gyrating ascent stage. For three unnerving minutes, the crew worked to stabilize the LM and gradually brought it under control.

During those three minutes, time both stood still and rushed by far too quickly for anyone on the ground to grapple with what was going on at the Moon. The radio signal went in and out as *Snoopy* pitched around, making communications and the data feed intermittent and spotty. While the crew ultimately solved the problem—putting their fighter-pilot skills to an incredible test—this episode highlighted for everyone, both on the ground and in space, that spaceflight was nowhere near becoming a routine experience. They could take nothing for granted.

When the two spacecraft came together after rendezvous and docking, Stafford radioed, "*Snoopy* and *Charlie Brown* are hugging each other." And the crew may have shared a few hugs among themselves too. They headed back to Earth, splashing down on May 26 in the Pacific Ocean.

"Apollo 10 was our last clearance test for the Apollo 11 lunar landing," said Glynn Lunney, "and basically the whole system—hardware and the people—passed the clearance test that we needed to be sure that we could go land on the Moon on the next flight."

The Apollo 10 Command and Service Modules (CSM) are photographed from the Lunar Module (LM) in lunar orbit. The high-gain antenna can be seen on the lower part of the CSM, which allowed for communications with Earth via the Unified S-Band system. Credit: NASA.

While Apollo 10 cleared the way for the first formal attempt at a lunar landing, everyone on the ground knew it would take a tremendous amount of additional training and planning to be ready.

IN 1969, THE SIMULATORS IN HOUSTON

and at the Cape were in almost constant use, with integrated sims for the flight controllers almost nonstop.

"If we weren't testing the crew, we were testing modifications to the spacecraft, to the flight software," said Frank Hughes. "The work was constant. That's why none of us were married at the time, or if you were, you weren't later. You were married to the job."

Meanwhile, the Simulation Branch at MSC remained on the lookout for any potential glitches in every system that could be possible malfunctions they could throw into a simulation.

"I'd go to the Mission Rules review meetings and listen for ideas for problems to put in the sims," said Jay Honeycutt. "One of astronaut Charlie Duke's responsibilities was to go to the same meetings, and we would sit side by side. Sometimes during conversations of certain issues, we'd whisper to one another, 'We should put this one in one of the sims.' And we'd jot them down."

Neil Armstrong participates in training for lunar surface operation in Building 9 in April 1969. Credit: NASA.

Dick Koos working as a SimSup, listening in on two communications loops at once. Image courtesy of Harold Miller.

Grumman and North American Rockwell had each supplied two people to the Simulation Branch who could go anywhere on the factory floors to get information from all the engineers about potential bugs or anomalies to help with ideas for the simulations. Jack Neubauer and Hank Otten were from Grumman, while Don Findlay and John Wills were from North American Rockwell.

At some point in early 1969, Neubauer came into the sim office and said to Dick Koos, "I asked the engineering guys about what could fail on the onboard computers, and they said, 'Oh, don't worry about it; it can't fail because it has rope memory.' But there has to be something that could give us a problem, right?"

Koos knew it would be hard to simulate a computer problem, because tinkering with the computations in the simulator might mess up the simulator programs.

"Why don't you just figure out how to get the simulator to do some type of erroneous event," Koos suggested, "or input a single event or signal into the computer program so it will give an error code?"

Neubauer said he'd look into it, but Koos didn't hear any more about it. And in the meantime, the simulation schedule and sequence of flights kept everyone on the sim team so busy, their lives were a blur. Koos thought about that conversation with Neubauer a few times, but he figured perhaps there wasn't a way to simulate a computer problem.

But Neubauer talked with Jack Garman in the Apollo Guidance Software Section at MSC and asked if he knew of a computer issue, something that might be totally software-related. Garman did a little research, jotted down a couple of semi-fatal computer errors (errors that would cause the computers to restart but not cause permanent damage) and told Neubauer how it might work to insert them in the simulation software.

July 5, 1969, was the day of the last scheduled set of integrated sims in Houston before the launch of Apollo 11. Armstrong, Aldrin and Collins were already at the Cape, so the astronauts in the LM simulator were the Apollo 12 backup crew: Dave Scott and Jim Irwin. Gene Kranz and his team of flight controllers were on the consoles in Mission Control and Koos was the SimSup.

Flight director Eugene F. Kranz is pictured during a simulation at the flight director console in Houston's Mission Control Center at the Manned Spacecraft Center site in 1965. Credit: NASA.

Jack Garman is second from left (wearing a jacket) in this 1969 view of the AGC Staff Support Room in Building 30 at MSC. Credit: NASA, image courtesy of Colin Mackellar.

"During the last afternoon session, a few of the guys came to me and said they wanted to put in this computer program alarm idea they had gotten from Neubauer," Koos said. "I said, 'Wait a minute, is that going to cause an abort?' We have a rule that the last simulation before the flight, everything should be fairly nominal as a confidence builder for everyone, so you can't put anything in that will cause an abort."

The rest of the team convinced Koos that Garman had assured them and provided documentation that the computer alarms they were inserting into the simulation were not serious enough to cause an abort.

A few minutes into the simulation run, Koos prompted the LM's computer to issue an alarm. Steve Bales, the guidance officer, was monitoring the LM's simulated computer telemetry and saw the caution and warning signal, listed as a 1201 alarm. Bales checked his manual and saw the 1201 alarm was an executive overflow, which meant the computer was overloaded. Garman, who was on station in the SSR and supporting Bales, knew this simulated alarm was a derivation of what he provided Neubauer, so he had to play both sides. He didn't offer any advice to Bales, and Bales didn't ask for any. No Mission Rules had been written on computer alarms because the flight controllers had never experienced any in a simulation before. Because the alarm kept sounding, Bales decided he needed to call an abort to the simulated Moon landing.

In the sim control room, Koos thought, *Oh, God*. He knew what was coming.

In the post-sim debriefing, Kranz was furious. First, he was mad at Koos for inserting an issue in the final sim that would cause an abort.

"We aborted, and I was really ready to kill Koos at the time, I was so damned mad," Kranz said. "We went into the debriefing, and all I wanted to do was get hold of him at the beer party afterward and tell him, 'This isn't the way we're supposed to train!'" Kranz was sure his flight control team had done everything right during the sim.

But, Koos explained, they had not done everything right. "You should not have aborted for those computer program alarms," he said. "What you should have done was taken a look at all of the functions. Was the guidance still working? Was the navigation still working? Were you still firing your jets? If so, ignore those alarms. And only if you see something else wrong in conjunction with that alarm should you start thinking about aborting."

APPLICABLE TO: IN DESCENT, AVERAGE-G ON

ALARM CODE		TYPE	PRE-MANUAL CAPABILITY	MANUAL CAPABILITY
00105	MK ROUT. BUSY	POODOO		
00430	CANT INTG. S.V.	"	PGNCS GUID. LOST,	PGNCS GUIDANCE NO/GO
01103	CCSHOLE- PROG. BUG	"		(PGNCS GO FOR
01204	NEG. WAITLIST	"	*PGNCS/AGS ABERT/ABRT STG	TAPE METER, CROSS-POINTERS,
01206	DSKY, TWO USERS	"		CONTROL
01302	NEG. SQ. ROOT	"	(decision how on	ABORTING)
01501	DSKY, PROG. BAD	"	current rules)	(NO LR DATA)
01502	DSKY, PROG. BUG	"	(NO LR DATA)	
00607	LAMB, NO SOLN	"		

"O.F." = Overflow, too many

CONTINUING OCCURRENCE OF:

01104	DELAY ROUT. O.V.	BAILOUT	DUTY CYCLE MAY DEGRADE PGNCS (AGS CONTROL MAY HELP- SEE BELOW)	SAME AS LEFT
01201	EXECT. O.F (VAC)	"	WATCH FOR OTHER CUES,	(except "other cues"
01202	EXECT. O.F. (JOBS)	"	PGNCS CONFIRM UNKNOWN,	which would otherwise
01203	EXECT. O.F (TASKS)	"	DSKY MAY BE LOCKED UP	be cause for ABORT
01207	EXECT. O.F. (MKS)	"	DUTY CYCLE MAY BE UP	PROBABLY AREN'T,
01210	TWO USERS	"	TO POINT OF MISSING SOME	INSTEAD IT WOULD
01211	MRK ROUT. INTRPT	"	FUNCTIONS (NAV, LAST TO DIE)	BE PGNCS GUIDANCE
02000	DAP O.F.	"	SWITCH TO AGS (FOLLOW ERR NEEDLES) MAY HELP (REDUCES PGNCS DUTY CYCLE SIGNIF.)	NO/GO — COMPLETES MANUAL LANDING IN AGS.)

ISS WARNING WITH:

00777	PIPA FAIL	LIGHT ONLY		
03777	CDU FAIL	"	PIPA/CDU/IMU FAIL	
04777	PIPA, CDU FAIL	"	DISCRETES PRESENT	same as left
07777	IMU FAIL	"		
10777	PIPA, IMU FAIL	"	(Other mission rules	
13777	CDU, IMU FAIL	"	suffice; alarm may help	
14777	PIPA, CDU, IMU FL	"	point to what rule will be broken)	

00214	IMU TURNED OFF	LIGHT ONLY	*AGS ABRT/ABRT STAGE	SWITCH TO AGS PGNS NO/GO on GOOD C (POSS. NO/GO on NAV.)
01107	E-MEM. DESTROYED	FRESH STRT	*AGS ABRT/ABRT STAGE	SWITCH TO AGS -PGNCS NO/GO! (IMU as ref. okay)

CONTINUING ←

| 00402 | BAD GUID. CMDS | LIGHT ONLY | *IF ALARM DOESN'T STOP: Same as POODO's ("ABRT.") | If ALARM DOESN'T STOP: Same as "POODO's" |

CONTINUING ←

| 01406 | GUID. NO SOLN | LIGHT ONLY | PGNCS GUID. NO/GO AS LONG AS ALARM OCCURRING (ATT. HOLD, CONST. GTC) CONT. OK) (ABRT WILL PROB. COME FROM CURRENT RULES eg. GTC vs. V) WATCH GTC ← | Same as left (except prob. no abort.) |
| 01410 | GUID. O.V. | | | |

Jack Garman's handwritten "cheat sheet" of computer alarms. Image courtesy of Colin Mackellar.

An engineering test version of the Apollo 11 US Flag, with Tom Moser, design engineer, Apollo Subsystem manager for Structures and Materials. Credit: NASA.

After Kranz realized Koos was correct, then he was furious with his team for not knowing what to do with these alarms, literally two weeks from launch. Specifically, he told Bales to work with Garman to study every computer alarm that existed in the code, from alarms that were normal to ones that could not possibly happen. And he didn't give a damn how long it might take—if they had to work all night or all week or every day from that moment until the launch, they needed to understand all the program alarms and know instantly what to do if this type of issue cropped up.

Bales and Garman conferred and wrote down all the alarm codes on a sheet of grid paper, with crib notes on what they meant and what their response should be. They each stuck the sheet under the plexiglass of their consoles. And then they asked Koos for more sims.

JUST TWO WEEKS BEFORE THE FLIGHT

of Apollo 11, Tom Moser was working late one evening in the Structures and Mechanics Division at MSC when his boss, Joe Kotanchik, entered Moser's office and shut the door.

"I'm going to give you an assignment, but you can't tell anyone about it, not your coworkers or your family," Kotanchik said to Moser, pausing to make sure the significance of what he was about to say was understood. "I need you to figure out how you can put a US flag and a mast somewhere on the LM so Neil and Buzz can put it on the Moon."

Moser knew that NASA had not been planning to place a US flag on the Moon. In 1967, the UN adopted the Outer Space Treaty, which stated that any planetary body or region of space could not be claimed by any nation. In the opinion of many, this effectively ruled out placing a flag on the lunar surface. But about three months prior to Moser's assignment, a few members of Congress approached NASA administrator Tom Paine, sharing their opinion that the Apollo 11 astronauts should place a US flag on the Moon to visibly show that this historic achievement had been accomplished by the United States. But it needed to be done in a way that wouldn't violate the UN treaty. Paine appointed an internal committee to come up with a plan, and they recommended that if NASA were to place a US flag on the lunar surface, there should also be a plaque bearing an inscription:

A team from MSC packs the US flag prior to the Apollo 11 mission in Building 9, Technical Service Shop. Left to right, unknown, Tom Moser (design engineer, Apollo Subsystem manager for Structures and Materials), Billy Druman (Technical Services technician), T. McGraw (deputy chief, Technical Services Division) and Jack Kinzler (chief, Technical Services Division). Caption courtesy of Tom Moser. Credit: NASA.

A technician holds the commemorative plaque that was later attached to the leg of the Lunar Module Eagle, engraved with the following words: "Here men from the planet Earth first set foot upon the Moon July 1969, A.D. We came in peace for all of mankind." It bears the signatures of the Apollo 11 astronauts Neil A. Armstrong, commander; Michael Collins, CM pilot; and Edwin E. Aldrin Jr., LM pilot, along with the signature of the US president Richard M. Nixon. Credit: NASA.

Here men from the planet Earth first set foot upon the Moon

July 1969, A.D.

We came in peace for all mankind

Additionally, the committee recommended Apollo 11 should carry small flags from each of the fifty US states and all member countries of the UN to be presented to each entity after the flight.

Besides carrying out the task in secret, Kotanchik gave Moser a few parameters.

"Joe told me there was no room in the Command Module or Lunar Module, so the flag and mast would have to be attached somewhere outside the LM," said Moser, "and the astronauts needed to be able to reach it and deploy it easily while wearing their spacesuit gloves. I also had to figure out how far they needed to stick the mast in the lunar surface and how far away from the Lunar Module the astronauts should put the flag so that it didn't burn up or blow over when they lifted off from the Moon."

Moser also needed to the assure the flag assembly could withstand the extreme environments of the launch, flight and lunar landing. He worked with three people in MSC's Technical Services Division—Jack Kinzler, Dave McCraw and Billy Drummond—and they developed and tested the concept.

Kinzler drew up a design for a specially designed, lightweight, telescoping flagpole that included a horizontal crossbar to hold the flag out fully unfurled to compensate for the lack of atmosphere on the Moon. It deployed similar to an umbrella, with a hinge on the horizontal bar.

"I was told someone went to different local department and hardware stores and bought plain old, regular 3-by-5-foot US flags and cut all the tags off so no company or store could claim it was their flag," Moser said.

The flags each cost about $5.50. A seam was sewn along the top of the flag where the horizontal bar could be inserted. The group made plans to put the flag assembly inside a protective shroud and attach it to the ladder of the LM. They wrapped the shroud in insulation so that it wouldn't be damaged by exhaust from the LM during landing.

Neil Armstrong, standing, and Buzz Aldrin, sitting, during spacesuit training at Kennedy Space Center. Credit: NASA.

Since Moser was a design engineer, he made sure the strength of the LM ladder wouldn't be compromised by attaching the shroud.

"Not being able to talk to anyone about this," he said, "I did all the stress analysis myself on the ladder with the flag assembly attached by putting it on a shake table, and shaking the hell out of it, with the loads it would experience during all the phases of flight."

The group erected and deployed the flag assembly to ensure it would operate properly on the Moon. Because the decision to include the flag and attach the plaque came so close to the launch date, George Low chartered a Learjet and flew Kinzler to Kennedy Space Center, where the flag assembly and the

commemorative plaque were installed in secret on Apollo 11's LM, out on the launchpad at 4:00 in the morning as the spacecraft sat atop its Saturn V rocket ready for launch.

The design and testing of the flag assembly had all happened within a few days. Realizing the significance of their work, Moser went back and retrieved a few of the scraps from the flag that had been cut off during its modification as momentos. And then he waited for the launch of Apollo 11.

EARLE KYLE WAS GOING TO THE APOLLO

11 launch. When the opportunity arose, he could hardly believe it. Kyle's family owned a small weekly African American newspaper in Minneapolis called the *Twin Cities Courier*, and somehow someone at NASA heard about Kyle's work on designing the Apollo hardware and asked his family's publication to send a representative to the Apollo 11 launch. The *Courier* was the only newspaper of its size and type in the country to get clearance for on-the-spot coverage of the Moon shot at the Cape.

Kyle knew he needed to capture the experience in the best way possible, not only for the newspaper but also for his own remembrance of this incredible opportunity. But he didn't have a good camera of his own, so he rented a Canon SLR from a local camera shop. As an engineer who tested everything, Kyle decided he needed to test out how to best take photos of the launch. He found the perfect setup right in front of him.

The Foshay Tower was the tallest building in Minneapolis in 1969, and it just happened to be about the same height as the Saturn V rocket as it sat on the launchpad. And the Honeywell plant where Kyle worked sat on a hill east of Minneapolis about 2.8 miles (4.5 km) away—as the crow flies—from the Foshay Tower. That was same distance between Launchpad 39A and the viewing stands at Kennedy Space Center where Kyle would be standing. During the week before he and his wife flew to Florida, Kyle used the Canon to test out various sun angles and camera settings, playing with the sensitivity and aperture settings, using up a few rolls of film just on the Foshay Tower. And all the while, he tried to imagine what it was going to be like to see the Saturn V start its journey to the Moon.

Right: The rollout of Apollo 11 Saturn V space vehicle from the Vehicle Assembly Building to Launch Complex 39A. Credit: NASA.

CHAPTER 9

APOLLO 11

There's no other way to classify what happened on Apollo 11's reentry except to say that we got lucky.

—GARY JOHNSON, NASA electrical systems and safety engineer

BEFORE DAWN ON JULY 16, 1969, EVERYONE began assembling at the Manned Spacecraft Center (MSC): the flight controllers and directors, as well as the engineers for the Mission Evaluation Room (MER), all the Staff Support Rooms (SSR), the Spacecraft Analysis Room, the auxiliary computer room and the Real Time Computer Complex (RTCC). With this being the fifth Apollo launch with a crew on board and the third mission to fly to the Moon, a hint of routine filled everyone's work. But today felt distinctly different.

Each of the missions leading up to Apollo 11 had their own unique characteristics: the successes and accomplishments, the problems in preparations, all the step-by-step processes that needed to be learned and mastered in simulations, the personalities of the crew and everyone involved. And the quick sequence of missions—five within nine months—meant there wasn't any time to bask in any successes. Instead, there was urgency and intensity.

"But in the whole course of it," Glynn Lunney said, reflecting on that morning, "the program had this energy that was pervasive. We had been involved in this whole thing for a long time—over eight years. There was a powerful sense of people wanting to pull off the Apollo Moon landing and return within the decade, of meeting the challenge and the goal."

The sense of intensity and excitement Lunney could feel that day was "like an electric field raising the hair on the body and stimulating the synapses firing inside the brain." He loved that feeling because, to him, it meant readiness and concentration.

And nobody wanted to screw up because July 16 was the launch day humanity had always dreamed about.

Left: The launch of Apollo 11 on July 16, 1969. Credit: NASA.

The early morning of July 16, 1969, found thousands of spectators on the beaches and roadways near the NASA Kennedy Space Center where they had camped the night before to witness history by watching the epic beginning of the journey of Apollo 11. Credit: NASA.

Earle Kyle stands at Kennedy Space Center, on hand for the launch of Apollo 11. Image courtesy of Earle Kyle.

THE DAY BEFORE APOLLO 11 WAS SCHEDULED

to lift off, Bob Wren and a few of his friends from MSC took off for Florida. "We had been working so hard, and we just decided at the last minute to go down there," Wren said, "and of course, we couldn't get a commercial flight. But one of the guys said, 'Heck, let's just fly ourselves down there!'"

Six friends piled into a Cessna Skywagon at the Houston airport and started their flight. But they had to fly around a thunderstorm, and the plane had mechanical problems—twice. They ended up making an emergency landing in northern Florida, then renting a car and barreling down the road among the hordes of traffic, barely arriving in time to see the launch.

EARLE KYLE HAD BEEN UP SINCE BEFORE

2:00 a.m. The bus that brought the media out to the Kennedy Space Center press site and launch viewing area left the hotel at an ungodly hour, and even at 2:00 in the morning, the air was uncomfortably warm. As the bus approached its destination, the view of Launchpad 39A looked like something out of a science fiction movie. The Saturn V rocket was bathed in floodlights, and the warm, humid air created a halo effect.

Now the morning sun was hot and bright, and the rocket gleamed as it sat on the pad, poised and peaceful. Humidity hung in the air and Kyle's eyeglasses and camera lens kept fogging up. On-site were thirty-five hundred members of the media and twenty-thousand VIPs, along with a million people packed onto the beaches in cars, campers and tents and parked alongside logjammed roads. They were all there to watch history. In the VIP viewing area, Kyle saw former president Lyndon B. Johnson and his wife, Lady Bird, Vice President Spiro Agnew, TV icon Johnny Carson and Ethiopian emperor Haile Selassie. Kyle even had the chance to talk with CBS News anchor Walter Cronkite.

Apollo 11's countdown proceeded smoothly, so Kyle made his way up to the top row of the bleachers; he wanted to stand above everyone else, with his back against the railing so he could have a good view of the launch and be able to hold his camera steady.

The launch of Apollo 11. Credit: Earle Kyle.

Excitement was in the air, but the majority in attendance had never witnessed the kind of power they were about to experience. Everyone watched the clock count down. At T minus nine seconds, the ignition sequence began, and with about six seconds to go the first flames snorted out of the F-1 engines. As the thrust built up, huge clamps held the rocket in place, but when the countdown clock hit zero, the clamps were released—and the 6.5-million-pound rocket began to climb, slowly, almost defiantly.

At the viewing site 2.8 miles (4.5 km) away, the light and heat arrived instantly; to Kyle, it felt like a blowtorch, but it was silent. He heard himself say, "Hey, I don't hear anything." But as the rocket slowly rose, the sound started to arrive, taking 14.7 seconds after ignition to travel across the water. The sound wave was almost visible, rolling across the lagoon, scattering the birds. And then the full sound hit, the rumble and the deafening roar, and it was like Mohammad Ali punching all the spectators in the gut, those five gigantic engines vibrating and banging like five cannons firing at the rate of a machine gun. It was all Kyle could do to hold the camera steady, but he got his pictures.

It took fifteen seconds after liftoff for the Saturn V to clear the tower, but the sound kept raining down on the crowd, giving them a long, loud show as the rocket picked up speed, rising faster on its journey. Kyle found that after the sound hit, the emotions hit. He started to cry, and hollered, "Hey, wait for me!" He knew he could never fly on a mission to the Moon, but he was honored that some of the equipment he had built and refined was on its way there.

Apollo 11 launch on July 16, 1969. Credit: NASA.

Inside the Spacecraft Analysis Room, called SPAN, at the Mission Control Center at MSC, during the Apollo 11 Mission. Credit: NASA.

As a vibration and acoustic expert, Bob Wren thought he knew what the Saturn V launch would be like. He knew that the sound wave from the F-1 engines is a frequency that couples with the natural resonate frequency of the human body. "Your whole body is coupling with that sound wave," he said, "and that's the reason why your whole abdomen and chest cavity starts shaking like crazy. It's not the sound coming through your ears, but the entire body just vibrating."

To Wren, the Saturn V appeared to just lumber its way off the pad, "You think, *Oh my God, is it ever going to clear the tower?*" he exclaimed. "It seemed like a week later it cleared the tower. But oh, what a sight, what a sight! To see something the size of a thirty-five-story building lifting off the pad . . . just amazing, just amazing. I'll never, ever forget it!"

DURING A BUSY TWO HOURS AND FORTY-four minutes after launch, the crew checked out all systems on the spacecraft, and after one and a half orbits around the Earth, the S-IVB stage reignited for a burn of nearly six minutes. Neil Armstrong, Buzz Aldrin and Mike Collins were on their way to the Moon. Collins detached the Command and Service Modules (CSM), *Columbia*, from the rocket stage. The Lunar Module (LM), *Eagle*, had ridden behind them, so Collins—in a nine-minute procedure—turned *Columbia* around and docked the spacecraft's nose to the docking port on the top of *Eagle*.

On board, Collins said *Columbia* "flies like a spacecraft instead of a simulator. Hope that's good."

About an hour later, after checking out the LM's systems, Collins "backed up," extracting *Eagle* from its holding spot in the spacecraft LM adapter, all while traveling at just under 25,000 miles per hour.

The crew finally had a moment to look out the window and saw Earth receding. "And Houston, you might be interested that out my left-hand window right now," reported Armstrong, "I can observe the entire continent of North America, Alaska, over the Pole, down to the Yucatan Peninsula, Cuba, northern part of South America, and then I run out of window."

Fourteen hours after lift-off, it was 10:30 p.m. back in Houston. A new shift across Mission Control and the support rooms now monitored all the systems. The three astronauts covered *Columbia*'s windows and went to sleep.

The next two days of the translunar flight were uneventful although no one felt flying to the Moon was at all routine. The crew sent live color TV transmissions to Earth: "Apollo 11 calling in from about 130,000 miles out," Armstrong said while showing views of Earth from space. He would later note that in his recollection of science fiction—from notable writers like Jules Verne and H. G. Wells—no writer had ever imagined that lunar explorers would be in constant communication with people back on Earth or, even more surprisingly, that transmissions and images would be shared in real time. Armstrong understood the importance of sharing their voyage, and his comprehensive descriptions of the views and their activities reflected that.

The rest of the flight to the Moon focused on keeping the spaceship operating smoothly. Mission Control was in constant contact with the crew, to assist them in tasks like making a midcourse correction, monitoring the fuel cells and making sure all the temperature and pressure readings stayed in the proper ranges. All the engineers in the MER observed and plotted their telemetry readings, feeling very much in touch with the systems, even across the ever-growing distances.

And the Moon grew larger out *Columbia*'s windows. On July 19, after traveling 240,000 miles in seventy-six hours, Apollo 11 flew behind the Moon out of contact with Earth for the first time since launch.

On the lunar far side, the crew fired the service propulsion system engine for exactly 357.5 seconds, slipping the spacecraft into orbit. They viewed the heavily cratered backside of the Moon with awe and argued over its color:

The Apollo 11 astronauts' view of the lunar farside, looking at a crater which is about 30 statute miles in diameter. Credit: NASA.

Spacecraft communicators are pictured as they keep in contact with the Apollo 11 astronauts during their lunar landing mission on July 20, 1969. From left to right are astronauts Charles M. Duke Jr., James A. Lovell Jr. and Fred W. Haise Jr. Credit: NASA.

Collins: "Look at those craters in a row . . . Something really peppered that one. There's a lot less variation in color than I would have thought, you know, looking down?"

Aldrin: "Yes, but when you look down, you say it's brownish color?"

Collins: "Sure." [Earlier, Collins had insisted the color was "plaster of Paris gray" but the Moon's color seemed to change with various sun angles.]

Aldrin: "Oh, golly, let me have that camera back. There's a huge, magnificent crater over here. I wish we had the other lens on, but God, that's a big beauty. You want to look at that guy, Neil?"

Armstrong: "Yes, I see him . . . What a spectacular view!"

Collins: "God, look at that Moon! . . . [That crater] is enormous! It's so big I can't even get it in the window. That's the biggest one you ever seen in your life. Neil? God, look at this central mountain peak."

The view of the Lunar Module Eagle, *as seen by Mike Collins in* Columbia. *Credit: NASA.*

When the spacecraft came around the Moon, the telemetry signal arrived on Earth right on schedule, and everyone in Houston knew the burn had been successful. After a few initial exchanges on the status of the spacecraft, Armstrong reported the view of the Moon looked very much like the pictures taken by the crews of Apollo 8 and Apollo 10, "but like the difference between watching a real football game and one on TV, there's no substitute for actually being here."

All was well in lunar orbit.

On July 20, everyone in Mission Control and all the support rooms felt the same tension and electricity as on launch day but magnified. Today, they would attempt to land on the Moon. The Mission Rules for Apollo 11 covered 218 pages, many of them detailing what actions were required for every contingency during the LM's lunar descent.

The seats in Mission Control were filled for this critical moment, with several astronauts joining Charlie Duke at the Capcom console or sitting where they could. More astronauts and nearly every high-ranking NASA official filed into the viewing room. The MER and SSRs swelled with people—everyone wanted to be on hand, even if they weren't scheduled to be on shift.

During the crew's eleventh orbit around the Moon, Armstrong and Aldrin moved to the LM to check out all the systems, confirming voice communications and telemetry links. Gene Kranz led his team of flight controllers in Mission Control, listening in on all the communication loops between the crew, the Capcom and all the backroom teams. Everyone told themselves this day should feel no different than the hundreds of hours spent in simulations and previous missions. But the difference was undeniable.

Ahead of the timeline, Armstrong and Aldrin continued their checkouts. On the twelfth orbit, at one hundred hours into the flight, *Eagle* undocked and separated from *Columbia*, with Collins performing a visual inspection, telling Armstrong and Aldrin they had a "fine-looking flying machine." Behind the Moon on the mission's thirteenth orbit, the LM descent engine fired for thirty seconds, providing the thrust necessary for slowing down for the first descent to the Moon's surface. Both *Columbia* and *Eagle* reappeared on the telemetry back to Earth as they came around from behind the Moon, and with another burn, Armstrong and Aldrin made their descent.

Unexpectedly, however, communication problems began: Radio communications drifted in and out from the LM; sometimes static filled the loop. Aldrin repositioned *Eagle*'s steerable high-gain antenna and communications returned but then dropped again as the spacecraft turned in flight. Kranz needed to make a decision: Keep going or abort? Telemetry showed all the spacecraft systems working well, however, and Kranz concluded they could press on, knowing in the back of his mind that his decision might need to change if communication problems persisted. Duke radioed up to Collins in *Columbia* to have him relay that *Eagle* was go for the powered descent initiative. They were truly going to land on the Moon.

Finally, communications seemed to stabilize as *Eagle* came closer to the Moon. Armstrong and Aldrin would be arriving at their landing site in about twelve minutes. All eyes in the Mission Control Center were on the consoles, keeping a close watch on the numbers, looking for any anomaly that could force an abort.

The Guidance and Navigation Computer on the LM was doing its job, and to this point, it ran in two-second cycles, reading all the parameters and performing calculations to pump out the next two seconds' worth of commands. Jack Garman described this type of navigation and flight control as if someone walking down the street opened their eyes once every two seconds to see their location and then decided how to proceed. "You see the hallway around you, and then you shut your eyes and then decide where to put your feet," he said. "If there are no obstacles ahead of you and you haven't reached the destination yet, take three steps before you open your eyes again, interpolate and figure out how many steps to take; and open your eyes, look around, shut your eyes, and go."

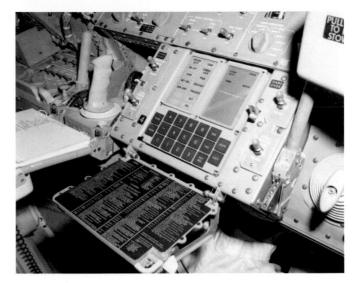

The Apollo Guidance computer on the Lunar Module. Credit: NASA.

But the closer the LM got to the lunar surface, the more precise the navigation needed to be. The computer would switch to one-second cycles, with a program called P63 running the computations for this phase of flight. The program held one of the busiest loads for the computer, certainly reaching—or perhaps exceeding—the 85 percent capacity requirements. But if nothing went wrong, the computer's executive should maintain its prioritized list of tasks.

Meanwhile, the spacecraft turned to make its final approach. A video camera positioned to look out Aldrin's side window allowed everyone back on Earth to see later what it looked like as *Eagle* came closer and closer to the lunar surface. Then Armstrong radioed to Houston that, based on their instruments and their view, they were going be "long," that they would overshoot their intended landing site. Suddenly, an alarm from the Caution and Warning System sounded.

"Program alarm," Armstrong called out with concern. He checked the display. "It's a 1202."

Steve Bales, the Guidance Officer in Mission Control, heard the call and saw it on his console. The alarm number sounded familiar. He quickly scanned the crib sheet of notes he and Garman had made after the simulation a few weeks earlier where he called the abort on the computer alarm. Before he could find it, however, Garman found it on his list and called Bales. The 1202 was an executive overflow, similar to the sim; the computer was overloaded, but it was designed to compensate. "As long as it doesn't reoccur, it's fine," Garman told Bales.

In the meantime, eighteen seconds had passed, and Armstrong urgently called for a reading for the 1202. Bales called, "We're go, Flight." In another twenty seconds, the alarm sounded again—another 1202. Aldrin called down, "Same alarm," noting that the alarm seemed to happen when they called up radar data to display the range to the landing site, along with the LM's velocity. As it turned out, this wasn't the cause of the alarms, but it was just more data for the computer to process. Aldrin asked Houston to call up the ranging data over the radio instead of the crew having to retrieve it from the computer.

But despite the alarms, the computer continued to handle the load of the landing program, and both Aldrin and Armstrong jubilantly called out when a critical scheduled throttle-down (a computer-driven reduction in speed from the thrusters) took place. "Wow, throttle down!" said Aldrin. "Better than the simulator!"

"Throttle down on time," Armstrong said decisively, now feeling confident in the computer again.

"You're looking great," Duke replied.

When the computer switched to the next phase, the landing approach program called P64, another alarm sounded, this time a 1201. "Same type!" Garman yelled to Bales. Then Garman heard the same call up the line, Bales to Duke, Duke to the crew: "Same type!" When the alarm sounded again, Garman yelled, "*Go, go, go,*" leaving no doubt on the question of continuing.

But then Armstrong noticed another problem. The landing program appeared to be sending *Eagle* toward a large crater, littered with enormous boulders. He entered the program named P66 on the computer, which gave him manual control of the LM. He pitched the craft forward to maintain enough speed to fly past the hazards, and after thirty more seconds, Armstrong eyed a suitable landing area and reduced *Eagle*'s speed to begin the final descent.

And that's when the low-fuel light came on. Only ninety seconds of fuel remained. It would now be a race against the clock.

Armstrong told Aldrin he found a good spot to land, while Aldrin continued to call out the landing parameters: "Still looks good. One hundred twenty feet. One hundred feet, 3½ down, 9 forward . . . Okay, 75 feet. And it's looking good; down a half. Six forward."

Just sixty seconds of fuel remained. Aldrin continued his call: "Forward, forward, 40 feet, down 2½. Picking up some dust."

At that moment, even amid the air of tension, everyone at the consoles in the Mission Control Center knew this landing was the real thing. In all the simulations and all the practice runs of landing on the Moon, they had never heard the words *picking up some dust*.

"Faint shadow, 4 forward. 4 forward. Drifting to the right a little. 20 feet, down a half," Aldrin reported.

Duke let the crew know they had just thirty seconds of fuel left.

About twelve seconds later, Aldrin called out, "Contact light," as the 3-foot (1-m) long probes on the end of the LM's landing footpad touched the lunar surface. "Okay. Engine stop . . . Descent engine command override, off. Engine arm, off."

After a few seconds of silence, Armstrong's steady, confident voice called up, "Houston, Tranquility Base here. The *Eagle* has landed."

Again, nothing in the simulations could have prepared everyone in Houston for that moment. Cheering erupted, but Kranz quickly brought all the teams under control, because several LM shutdown activities need to be performed and a series of "stay–no stay" decisions needed to be made within forty seconds. The astronauts and everyone in Mission Control needed to determine if there were any reasons the astronauts might need to make an emergency lift-off from the Moon. It would take time to arm the ascent engine, and there was only a four-minute window where they could lift off, catch and rendezvous with Collins in *Columbia*.

One of the first decisions was an "upright" call. Ken Young, Dave Alexander and Jerry Bell sat at the console in the vehicle systems Staff Support Room, supporting Flight Dynamics Officer Jay Greene, listening to Aldrin send down the initial readouts on the LM systems. When they heard the vertical alignment reading, Young, Alexander and Bell needed to determine if there was any danger of the LM tipping over, or if the angle could cause problems for lift-off (in case one of the landing pads had come down inside a small crater or depression or the lunar surface really was "fluffy" as one scientist had predicted).

"The tip angle we could tolerate was 15 degrees," said Young, "and in the backroom, when we got Buzz's callout, the angle was only about 2 degrees. It wasn't even close, but we checked our charts to make sure. It was a tense first few seconds until we got the call because we knew we only had about twenty seconds to make our decision."

The view of Tranquility Base out the window of the Eagle. *Credit: NASA.*

Other data points included the status of the life-support system, fuel readings for the ascent stage and communications. But soon, Mission Control advised Armstrong and Aldrin—in NASA parlance—they were "go" to stay.

But just after the four-minute ascent window had passed, a problem arose. A pressure reading in the fuel line leading to the descent engine was rising rapidly, with the temperature rising as well. Some of the fuel had frozen into a "plug" due to a surge of cold helium after the engine shut down, blocking the fuel line. Grumman engineers looked at the problem and knew if the temperature reached 400°F (204°C), the fuel could become unstable and explode. A procedure was quickly readied to instruct the crew to "burp" the line to release the pressure. But just as suddenly, the pressure and temperature dropped. The frozen plug had likely melted due to the rising temperature of the fuel. Everyone in Mission Control and at Grumman breathed a sigh of relief.

AFTER THE *EAGLE* HAD LANDED, EVERYONE involved with the Apollo Guidance and Navigation Computer collectively drew their breaths. Those in Boston at the Instrumentation Lab and the team members on-site at MSC needed to confer—quickly. Hardly anyone had heard of the 1202 and 1201 alarms, much less knew what they meant. Of course, Jack Garman knew the cause was an overload of data for the computer, but what caused the alarm to sound during Apollo 11's landing?

"As *Eagle* sat on the lunar surface the evening of July 20, you can imagine how all of us on the ground felt, as we had just seen our software puke up its guts during the landing," said Ken Goodwin, a Massachusetts Institute of Technology (MIT) system engineer who was stationed in the MER during Apollo 11. "We knew we had been saved by Hal Lanning's priority-driven executive and waitlist operating system, coupled with NASA's requirement to incorporate a restart capability into the landing software. But foremost in our minds, we knew the ascent program for leaving the moon was just as demanding computer-wise as the descent programs. We had just barely made it down to the lunar surface. How were we going to get back up again?"

A view of the Mission Evaluation Room during Apollo 11. Ken Goodwin is seated on the far right. Credit: NASA, image courtesy of Ken Goodwin.

MIT computer team at the Instrumentation Laboratory during the Apollo 11 mission. Facing the camera are Eldon Hall and Dick Battin. Credit: Draper.

The Guidance, Navigation and Control (GNC) team worked through the night to determine the problem. It turned out that the overload came from a combination of reasons, not just one: a late change to the configuration of the software and an underlying known issue with the computer's electronics that was thought to have less than a 1 percent chance of ever happening.

Over the years, blame has been placed on a "checklist error," the assumption being that Buzz Aldrin turned on the rendezvous radar when it wasn't needed. The rendezvous radar's job was to scan for the orbiting Command Module (CM) for a potential rendezvous. For landing, it normally wouldn't be required. But in the event of a landing abort, having the radar on and ready to go could have been beneficial in terms of saved time and greater efficiency.

Having the rendezvous radar switched to the Lunar Module guidance computer (LGC) was a change that Aldrin had asked for. This change had been analyzed, studied and approved by several people involved with the software, and it seemed innocuous at the time. It was added to the crew's checklist and timeline, so turning the radar on was not a crew mistake; Aldrin did exactly what the instructions told him. The problem came in that the change was made shortly before Apollo 11's flight and the procedure wasn't fully tested.

"What Buzz Aldrin thought—and a small group in the [Instrumentation] Lab concurred—is that it would save a little procedural time during a landing abort if the rendezvous radar was switched to the LGC versus the fully tested operational procedure of having it in the AUTO/SLEW switch configuration," said Goodwin.

This setting meant the radar had to be manually positioned by an astronaut and that the radar wouldn't send data to the computer. Unfortunately, what no one knew at the time was that the radar was inundating the computer with data, due to a combination of the procedural change in settings—which was made after most of the formal testing on the software had been completed—and the underlying problem with the computer's electronics.

"The classic software belief is 'don't change your software at the last minute.' Anyone in the software business knows that as a rule not to be violated, but that's exactly what we did," said Goodwin.

The underlying problem had to do with the phasing of two power supplies in the computer. The frequency between the two power supplies should have been frequency-locked and phase-synchronized. However, the original designs called for only frequency locking between the two. MIT member George Silver, who usually worked at Cape Canaveral, identified this as a potential problem before the Apollo 11 flight, but the Instrumentation Lab's software engineers determined the chance of it causing a substantial issue was extremely small. With a short timeline and a low probability of failure, the software engineers decided to fly Apollo 11's power supplies as they were, and the fix would be instituted for subsequent flights.

Unfortunately, the rendezvous radar was turned on at the exact moment where the phase synchronization would cause a problem. The computer couldn't make sense of the radar settings, causing constant interruptions to the computer.

"Only in the LGC switch configuration would the underlying problem of the power supplies' phase synchronization become a problem that had a 1 percent probability of occurring, based on when the computer was turned on," said Goodwin.

As luck would have it, the power "on" sequence hit that 1 percent sweet spot that caused the rendezvous radar to overburden the computer. "When the computer dropped from a two-second cycle to a one-second cycle," Garman explained, "suddenly there was this extra involuntary load on the computer, meaninglessly adding data to the cycles. It needed to run at over 100 percent capacity, meaning there wasn't enough time to do everything—hence the problem."

"George Silver realized what had happened," Goodwin said, "and he got through to us in Mission Control, allowing us to get to the root cause. We were able to come up with a few workarounds to avoid the computer being overloaded on ascent and we had them instituted about eight hours before the astronauts would lift off from the Moon. We didn't get any sleep, but knew we would avoid the problem during Apollo 11's lift-off from the Moon."

Solving this problem for the subsequent Apollo missions would require changes to the hardware. But could something else be hidden among the components that might cause the computer to overload on the next flight? "That started a search for problems that went on for years," Garman said.

The serendipity of the computer alarm simulation that included Garman and SimSup Dick Koos can hardly be overstated. That essential and fortuitous training just fifteen days prior to the Apollo 11 Moon landing very likely changed the course of history.

At the Honeysuckle Creek Tracking Station during the Apollo 11 mission. Standing is Mike Dinn, deputy station director. Image courtesy of Colin Mackeller.

WHEN NASA OFFICIALS MADE THE decision to move up the timeline for the extravehicular activity (EVA)—the moonwalk—Mike Dinn was ready. Dinn was the deputy station director of the Honeysuckle Creek tracking station near Canberra, Australia, part of NASA's Manned Space Flight Network of radio dishes that made communications for Apollo possible. Three 85-foot (26-m) antennas equally spaced around the world were located at Goldstone, California; Fresnedillas, Spain; and Honeysuckle Creek.

With the timing of the moonwalk, Honeysuckle Creek was the prime station assigned to receive the initial TV pictures from the Moon. Dinn had trained his team at the station with simulations, similar to how NASA trained astronauts.

"I had been in Houston for a meeting," Dinn said, "and had the chance to take part in the simulations in Mission Control, where they were simulating communications and switching to the various tracking stations. I was so impressed that when I returned to Australia, I instituted our own simulations. We trained and practiced and gave a lot of thought to our contingency plans for any possible failures."

Dinn strived to make his team's simulations as realistic as possible and even tried to find someone who spoke with an American accent to play the role of the astronauts; but alas, one of their own team had to play the part.

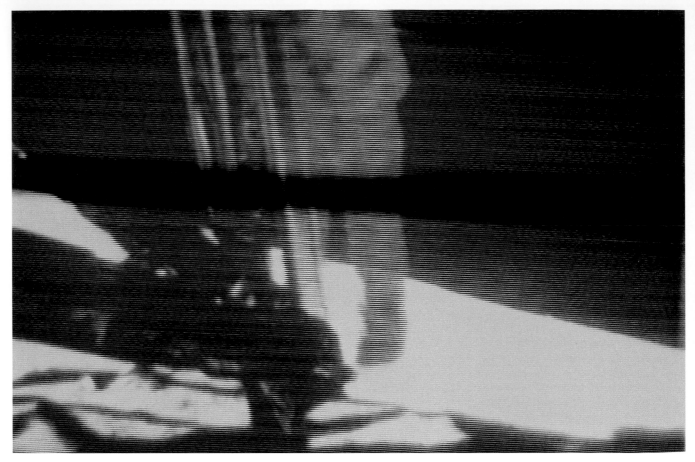

The televised view of astronaut Neil Armstrong descending the ladder of the Apollo 11 Lunar Module (LM) prior to making his first step on the Moon. Credit: NASA.

For television footage to be broadcast live from the Moon took several levels of coordination. NASA determined a heavy-duty camera used by the Department of Defense was capable of withstanding the rigors of space travel. The camera was stowed in just the right spot inside the LM's Modularized Equipment Stowage Assembly (MESA), which held the equipment and experiments the astronauts needed for their EVA. As Armstrong stepped onto the first step of the LM, he pulled a release to open the MESA, allowing the camera to peek out from its insulated perch, just to the left of the LM's ladder. Aldrin activated a circuit breaker inside the LM, turning the camera on. Armstrong's first steps could now be visible to the estimated six hundred million people around the world watching this historic event on their televisions.

Back in Houston, Tom Moser—who had constructed the assembly for the US flag—watched tensely as Armstrong stepped on each rung. When Armstrong leaped from the last step, in a flash, Moser visualized that the ladder had failed, and a jagged edge of metal might penetrate the astronaut's pressurized spacesuit—all because of the last-minute decision to add the flag to the flight. But all was well. Armstrong just needed to jump down to the Moon. He had landed *Eagle* so gently that its shock absorbers didn't compress as designed, making the ladder about 3 feet (1 m) higher above the lunar surface than expected.

At 9:56 p.m. Central time on Sunday, July 20, Armstrong first set foot on the Moon, saying, "That's one small step for a man, one giant leap for mankind."

One of the only images of Neil Armstrong on the Moon during the historic first EVA on the Moon. Armstrong is standing at the modular equipment storage assembly (MESA) of the Lunar Module. Credit: NASA.

Immediately Armstrong began to share what he was experiencing: The nature of the lunar surface was a fine, powdery material. He noted he sunk in only ¼ inch (0.5 cm) or less and that the LM footpads had penetrated only a few inches. He observed that the exhaust of the descent engine had not cratered the area directly below the LM engine nozzle. Like a tourist, Armstrong took pictures and—after some prodding from Mission Control—collected the first lunar samples, which he stowed in a spacesuit pocket.

Aldrin soon followed him out, and the camera was positioned on a tripod about 30 feet (9 m) from the LM. People on Earth watched in wonder as the two astronauts quickly learned how to skip around the landing site in the one-sixth lunar gravity, collecting rocks and soil samples and setting up their experiments, the Apollo Scientific Experiments Package. When they prepared to place the US flag on the lunar surface, the telescoping top support bar would not fully deploy. This gave the partially extended flag the appearance of waving in the breeze. The astronauts talked with President Nixon, then placed commemorative medallions on the Moon's surface that bore the names of the three Apollo 1 astronauts and two cosmonauts who gave their lives in the pursuit of space exploration. A 1½-inch (4-cm) silicon disk containing miniaturized goodwill messages from seventy-three countries also stayed behind.

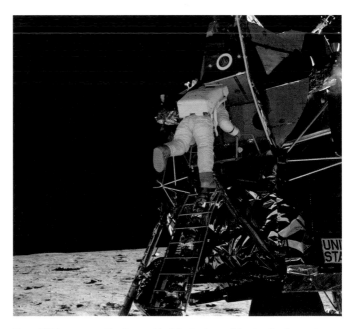

Buzz Aldrin egresses the Lunar Module Eagle *and begins to descend the steps of the LM ladder as he prepares to walk on the Moon. This photograph was taken by astronaut Neil Armstrong during the Apollo 11 extravehicular activity (EVA). Credit: NASA.*

The view of Earth from lunar orbit, as seen by the astronauts of Apollo 11. Credit: NASA.

A view of Mission Control in Houston during the Apollo 11 moonwalk. Credit: NASA.

The entire EVA lasted more than two and a half hours. Norman Chaffee watched from home with his family, Jerry Woodfill took his shift in the MER, Frank Hughes watched from Mission Control, Jerry Bell and Ken Young stayed at MSC even though their and shift had ended. No one wanted to miss the moment they worked so hard to make possible.

Armstrong and Aldrin returned to the LM and tried to sleep. Overnight, engineers worked in Houston to solve the two problems that weighed heavy on everyone's minds: Where was *Eagle* on the lunar surface, and would the broken switch for the ascent engine work?

There was only one way to figure out their location. The Capcom woke Aldrin early to perform a rendezvous radar check. The vectors between the LM and the CM in orbit allowed engineers in Mission Control to determine the landing site. It was about 5 miles (8 km) away from any spot they had pinpointed earlier.

As Armstrong, Aldrin and Collins prepared for lift-off and rendezvous, engineers at MSC devised a potential workaround for the broken ascent engine switch. But Aldrin came up with his own solution. He used his felt-tip Duro Pen marker and pushed in the breaker. The circuit became live, and the crew was now ready to proceed with the countdown to leave the Moon.

In what many considered one of the riskiest moments of the entire mission, Armstrong and Aldrin activated the ascent engine, while explosive bolts separated the top stage from the bottom descent stage. It all worked perfectly, and the ascent stage zoomed to orbit to meet with *Columbia* for docking. The astronauts transferred the lunar samples and film canisters to *Columbia*, performed the required maneuvers to prepare for leaving the Moon and headed for home.

Right: The view out the window of the Lunar Module after the first EVA on the Moon. The flag and the astronauts' footprints are visible on the lunar surface. Credit: NASA.

Apollo 11 astronauts wearing their Biological Isolation Garments (BIGs) after splashdown and recovery. Credit: NASA.

The Apollo 11 astronauts wearing their Biological Isolation Garments (BIGs) as they emerged from the helicopter that brought them on board the USS Hornet. *Credit: NASA.*

IN THE PREDAWN DARKNESS ON JULY 24, *Columbia* splashed down 920 miles (1,480 km) southwest of Honolulu, 13 miles (21 km) from the prime recovery ship, the USS *Hornet*, giving Apollo 11 a mission duration of eight days, three hours, eighteen minutes and thirty-five seconds. Recovery crews arrived quickly via helicopter and the Navy frogmen tossed packages with the Biological Isolation Garments (BIGs) into the spacecraft. To avoid "infecting" the largest body of water in the world with any potential Moon bugs, the recovery crew sprayed a decontaminate over the CM, and the astronauts were each wiped down with a sodium hypochlorite solution after they got into the life raft. They were each lifted aboard the helicopter and whisked to the *Hornet*.

Looking like space aliens in their BIG garb, Armstrong, Aldrin and Collins emerged from the helicopter, waving heartily to the surrounding ship's complement—even though they couldn't see much through the fogged-over gas mask visors.

They walked quickly to the Mobile Quarantine Facility. All three wanted out of those BIGs as soon as possible: It was a warm day, and inside the helicopter, they had started to heat up inside the rubber suits. "I can remember thinking, *I'm giving these guys thirty seconds, and then I'm getting out of this goddamned suit*," Collins said later. "I didn't care how many bugs were coming with me. But I wasn't ticked off; I was very pleased."

Meanwhile, recovery crews brought *Columbia* on board the *Hornet* and connected it to a special containment area where all the materials from the interior of the craft (such as all the exposed film and sample return containers) were taken into the quarantine trailer then passed through a decontamination lock. The items were immediately flown to the nearest airport, located on a small atoll called Johnston Island, then flown to Houston. The Apollo 11 crew spent five days secluded in their Mobile Quarantine Facility as the *Hornet* transported them to Pearl Harbor, Hawaii.

Around the country and the world—but especially in Houston—the celebrations began as soon as *Columbia* hit the water. In Mission Control, the MER and all the support rooms, cheers, handshakes and celebratory cigars abounded, along with the words, "We did it!" Thoughts turned to everything that had happened along the way: the hard work and long hours, the sacrifices, the fun, the camaraderie, the Apollo 1 fire and the rebuilding from tragedy. Glynn Lunney looked around at all the young faces—the average age of the flight control team in 1969 was twenty-eight. They had done something that started out as impossible, Lunney realized, and it was accomplished in eight years and two months from President Kennedy's speech in May 1961.

Celebrations in Mission Control after Apollo 11's splashdown. Credit: NASA.

It took several hours for everyone to wrap up their work and button up all the data from the mission. But by the time 5:00 p.m. rolled around, NASA Parkway was a complete gridlock. Everyone from MSC was heading to the bars and restaurants to celebrate, and all of Houston came to join them. Soon people just parked their cars along the side of the road and walked so they could join in at one of the now-legendary splashdown parties. By morning, many would wonder how they got home or where they might find their car.

At about 6:00 p.m., Henry Pohl's wife threaded her way through the parked cars and crowds to pick up her husband. "We just had to poke along because the streets were just solid, wall-to-wall people," Pohl said. "Everybody was celebrating. I didn't go to any celebrations, though. My wife had the car loaded and we took the kids out to the country and relaxed for a while. It had been a long, long stretch with not much sleep. Matter of fact, I lay down in the back of the station wagon and went to sleep."

Max Faget with an early model of the Space Shuttle. Credit: NASA.

EPILOGUE

APOLLO 12–17

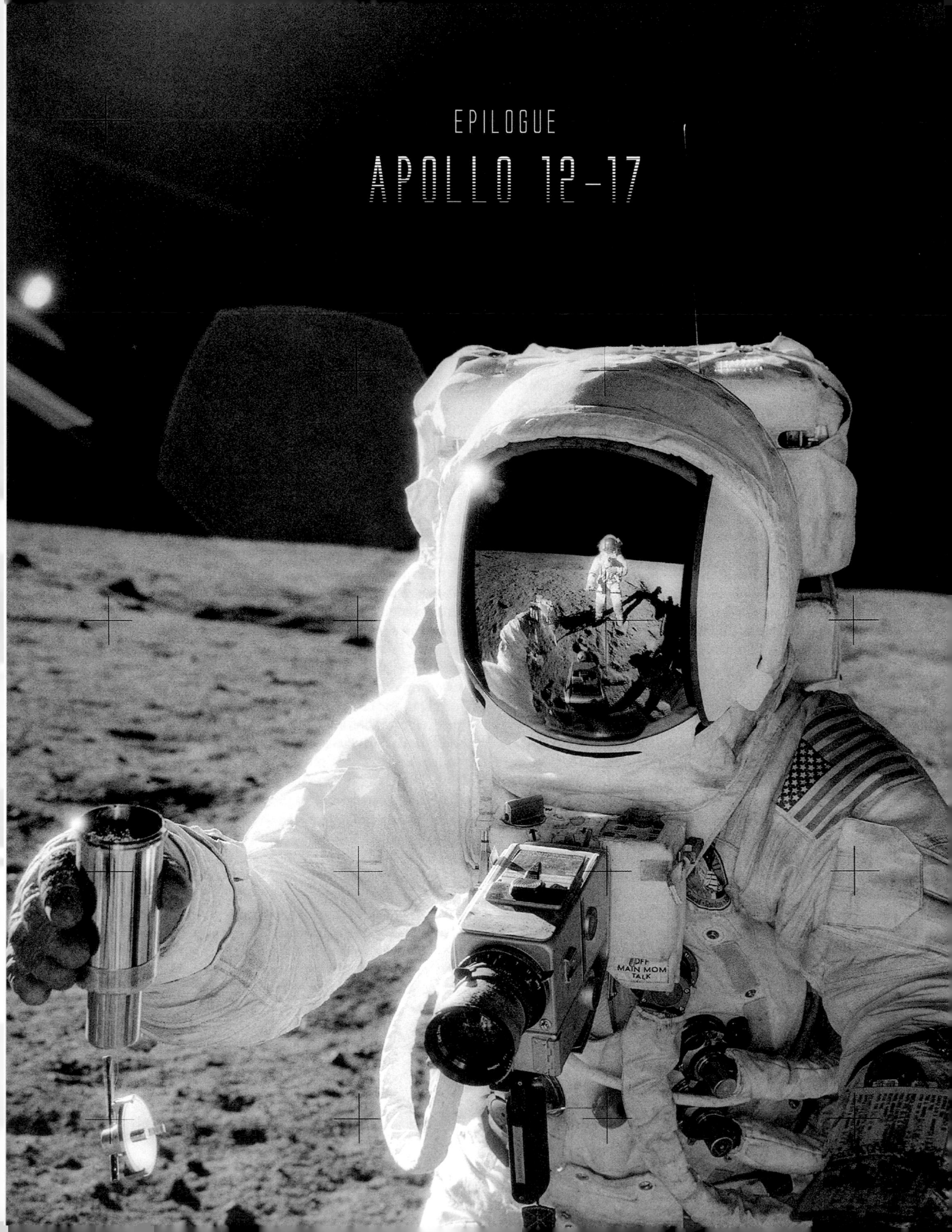

APOLLO 12

Apollo 12 launched successfully from Kennedy Space Center on November 14, 1969, with the crew of Pete Conrad, Alan Bean and Dick Gordon. But the launch wasn't without a little drama. Thirty-six-and-a-half seconds after lift-off, the rocket was struck by lightning. Sixteen seconds later, a second bolt of lightning rattled the vehicle, taking out much of the spacecraft's electrical system. Quick thinking by flight controller John Aaron in Mission Control had Bean flipping a re-set switch, and Apollo 12 continued on its way into orbit, and a full recovery meant the mission could continue to the Moon.

Since Apollo 11 had landed so far from its original target, the crew of Apollo 12 wanted to prove they could do a pinpoint landing. The LM Intrepid came down 600 feet (183 m) from the 1967 Surveyor 3 spacecraft, and the crew brought back Surveyor's camera. Conrad and Bean collected lunar rocks and deployed the Apollo Lunar Surface Experiments Package, or ALSEP, to gather seismic, scientific and engineering data. The three astronauts safely returned home, splashing down on November 24, 1969.

Top left: The three astronauts for the second lunar landing mission, Apollo 12, left to right, are Charles (Pete) Conrad Jr., Richard F. Gordon Jr. and Alan L. Bean. Credit: NASA.

Right: The huge, 363-foot (111-m) tall Apollo 12 Saturn V launched from Launch Complex 39 at Kennedy Space Center on November 14, 1969. Credit: NASA.

Bottom left: This unusual photograph, taken during the second Apollo 12 moonwalk, shows two US spacecraft on the surface of the Moon. The Apollo 12 Lunar Module (LM) is in the background. The unmanned Surveyor 3 spacecraft is in the foreground. The television camera and several other pieces were taken from Surveyor 3 and brought back to Earth for scientific examination. Here, Conrad examines the Surveyor's TV camera prior to detaching it. Credit: NASA.

Facing page: Astronaut Alan Bean holds a Special Environmental Sample Container filled with lunar soil collected during the extravehicular activity (EVA) where Bean and Conrad walked on the lunar surface for nearly four hours. Conrad, who took this picture, is reflected in Bean's helmet visor. Credit: NASA.

APOLLO 13

Apollo 13 was supposed to be NASA's third mission to land on the Moon. But during the third day of the trip to the Moon for astronauts Jim Lovell, Fred Haise and Jack Swigert—when the crew was nearly 200,000 miles from Earth—an oxygen tank in the Service Module blew up. And so began the most perilous but eventually triumphant situation ever encountered in human spaceflight.

The crew "powered down" the crippled Command Module, and moved into the Lunar Module, normally designed to be used only for landing on the Moon. The LM now served as a "lifeboat" for the crew, providing systems for power, propulsion and life support.

Engineers in the MER and other mission support rooms were instrumental in figuring out work-arounds (such as fitting the square CM lithium hydroxide canisters in the round hole of the LM's system to remove carbon dioxide from the air inside the spacecraft) and computing the exact burns needed to get the spacecraft on the correct trajectory back to Earth. Despite the odds, the crew returned safely to Earth, initially thinking they had failed. But instead, the mission has become the prime example of NASA's ingenuity and their ability to solve problems on the fly.

Top left: A group of flight controllers gather around the console of Glynn Lunney (seated, center) along with several NASA/MSC officials during Apollo 13. Credit: NASA.

Top right: Astronaut Jim Lovell pictured at his position in the Lunar Module (LM) during the flight of Apollo 13. Credit: NASA.

Bottom left: Interior view of the Apollo 13 Lunar Module (LM) during the trouble-plagued journey back to Earth. This photograph shows some of the temporary hose connections and apparatus which were necessary when the three astronauts moved from the Command Module to use the LM as a "lifeboat." Astronaut John (Jack) Swigert Jr., Command Module pilot, is on the right. The astronauts needed to improvise to use the Command Module lithium hydroxide canisters to purge carbon dioxide from the LM. Credit: NASA.

Bottom right: This view of the severely damaged Apollo 13 Service Module (SM) was photographed from the Command Module following SM jettison. An entire panel on the SM was blown away by the explosion of an oxygen tank. Credit: NASA.

APOLLO 14

Apollo 14 launched on January 31, 1971, just a four-month slip in the schedule after dealing with the issues encountered during Apollo 13. Commander Alan Shepard and Lunar Module Pilot Ed Mitchell landed at the Fra Mauro region, while Command Module pilot Stuart Roosa remained in orbit. The astronauts used the Modularized Equipment Transporter (MET), a small cart, to haul equipment during their two EVAs. They collected samples, took photographs and went on an extended geological traverse to nearby Cone crater. While scientists were excited for the geological findings of the mission, the public's foremost recollection seems to be Shepard hitting two golf balls on the Moon, near the end of the second moonwalk. The crew returned to Earth on February 9, 1971.

Top left: The prime crew of the Apollo 14 lunar landing mission. Left to right are Edgar D. Mitchell, Alan B. Shepard Jr. and Stuart A. Roosa. The Apollo 14 emblem is in the background. Credit: NASA.

Top right: A front view of the Apollo 14 Lunar Module (LM), which reflects a circular flare caused by the brilliant sun, as seen by the two Moon-exploring astronauts (out of frame) of the Apollo 14 mission during their first EVA. In the left background is Cone Crater. In the left foreground are the erectable S-Band antenna and the US flag. Credit: NASA.

Bottom left: Astronaut Alan Shepard stands beside a large boulder on the lunar surface during the mission's second moonwalk on February 6, 1971. Note the lunar dust clinging to Shepard's space suit. Credit: NASA.

Bottom right: The Apollo 14 Lunar Module (LM) ascent stage lifts off the lunar surface and the powerful LM engine causes a brief force of wind, which scatters some of the gold-colored foil that covers the LM and also disturbs the US flag. This picture was taken from a camera mounted inside the LM. Credit: NASA.

APOLLO 15

Apollo 15 launched on July 26, 1971, with the crew of Dave Scott, commander; Al Worden, Command Module pilot; and Jim Irwin, Lunar Module pilot. This mission saw the first use of the "moon buggy"—the Lunar Roving Vehicle (LRV) which astronauts used to explore the geology of their landing site, the Hadley Rille/Apennine region. The LRV allowed the Apollo 15 crew to venture a total of about 18 miles (29 km) during their three moonwalks, driving at speeds as high as 10 m.p.h. Scott and Irwin collected samples from the low dark plains, the Apennine highlands and the area along Hadley Rille, a long, narrow winding valley. After leaving the lunar surface and docking with the CM, flight surgeons noticed some irregularities in Irwin's heartbeat, delaying the jettison of the LM and other activities, but Irwin's heartbeat returned to normal, allowing the mission to proceed. Worden conducted a 33-minute stand-up EVA on the return trip to Earth, to collect film canisters from outside the CM, and the crew of Apollo 15 splashed down on August 7, 1971.

Top left: The crew of Apollo 15, left to right, David R. Scott, Alfred M. Worden and James B. Irwin. The crew is posed behind the subsatellite that they later deployed from lunar orbit. Credit: NASA.

Top right: Dave Scott is seated in the LRV during the first Apollo 15 moonwalk. This photograph was taken by Jim Irwin. Credit: NASA.

Bottom left: Astronaut Jim Irwin gives a military salute while standing beside the deployed US flag on the lunar surface at the Hadley-Apennine landing site during the mission's second EVA. The LM Falcon is in the center. On the right is the Lunar Roving Vehicle (LRV). Hadley Delta in the background rises approximately 13,124 feet (4 km) above the plain. The base of the mountain is approximately 3 miles (4.8 km) away. This photograph was taken by astronaut Dave Scott. Credit: NASA.

Bottom right: The LM Falcon is seen against the barren lunarscape during the third Apollo 15 EVA. The object next to the US flag is the Solar Wind Composition (SWC) experiment. Note bootprints and tracks of the LRV. The light spherical object at the top is a reflection in the lens of the camera. Credit: NASA.

APOLLO 16

The Apollo 16 crew of John Young, commander; Ken Mattingly, CM pilot; and Charlie Duke, LM pilot, launched on April 16, 1972. This was the second of three science-oriented missions for the Apollo program, with the primary objective of investigating the Descartes highlands area on the Moon, considered to be representative of much of the Moon's surface. During the first EVA, the crew deployed a nuclear-powered, automatic scientific station called Apollo Lunar Surface Experiment Package (ALSEP), and the second and third EVAs were devoted to geological exploration and sample gathering using the LRV. The crew conducted a total of 20 hours and 14 minutes of lunar surface activities and collected over 700 individual pieces of lunar rocks and soil. Young and Duke traveled over 16 miles (26 km) in the LRV, staying on the lunar surface for 71 hours. The crew concluded their mission on April 27, 1972.

Top left: The launch of Apollo 16 on April 16, 1972. While astronauts Young and Duke descended in the LM Orion to explore the Descartes highlands region of the Moon, astronaut Mattingly remained with the CSM Casper in lunar orbit. Credit: NASA.

Top right: A partial view of the Apollo 16 Apollo Lunar Surface Experiments Package (ALSEP) in deployed configuration on the lunar surface. The Passive Seismic Experiment (PSE) is in the foreground center; Central Station (C/S) is in center background, with the Radioisotope Thermoelectric Generator (RTG) to the left. One of the anchor flags for the Active Seismic Experiment (ASE) is at right. Credit: NASA.

Bottom left: Charlie Duke collecting lunar samples at Station No. 1, during the first Apollo 16 EVA. Duke is standing at the rim of Plum crater. The parked LRV is in the background, left. Credit: NASA.

Bottom right: The LM Orion and the LRV as seen during the first Apollo 16 EVA. The lunar surface feature in the left background is Stone Mountain. Credit: NASA.

APOLLO 17

Apollo 17 was the final mission to land on the Moon and included the only trained geologist to walk on the lunar surface, Lunar Module pilot Harrison Schmitt. He was joined by Gene Cernan, commander, and Ron Evans, CM pilot. They launched just before midnight of December 7, 1972, the only night launch of the Apollo program.

Apollo 17 astronauts traveled the greatest distance (21 miles [34 km]) on their lunar EVAs with the LRV and returned the largest amount of rock and soil samples (238 pounds [108 kg]) from their landing site, a lunar valley called Taurus-Littrow. Apollo 17's astronauts spent a record 22 hours performing EVAs, and Cernan and Schmitt performed the first "car repair" (with duct tape) on the Moon after a fender broke on the LRV. The astronauts deployed several scientific instruments and used a traverse gravimeter, an instrument they carried along on the Rover, measuring the relative gravity at several locations, providing information about the lunar substructure. Before Cernan climbed aboard the LM Challenger for the last time, he gave a short speech, that "America's challenge of today has forged man's destiny of tomorrow. And, as we leave the Moon at Taurus-Littrow, we leave as we came and, God willing, as we shall return: with peace and hope for all mankind."

The crew splashed down on December 19, 1972, in the South Pacific Ocean, and Cernan still holds the distinction of being the last person to walk on the Moon.

Top left: The crew of Apollo 17, the last lunar landing mission: Eugene A. Cernan (seated), commander; Ronald E. Evans (standing on right), Command Module pilot; and Harrison H. Schmitt, Lunar Module pilot. They are photographed with an LRV training vehicle. Credit: NASA.

Right: Harrison Schmitt working beside a huge boulder during the third Apollo 17 EVA. Credit: NASA.

Bottom left: Wearing special germ-free clothing, Dr. Robert R. Gilruth, right, inspects lunar samples collected during the Apollo 17 mission, NASA's final Apollo flight. Credit: NASA.

Facing page: Gene Cernan walks toward the LRV during the moonwalk at the Taurus-Littrow landing site. The photograph was taken by Harrison Schmitt. Credit: NASA.

Tylko, John (chief innovation officer, vice president of innovation and director, Aurora Flight Sciences)

Vaughan, Chester (Propulsion and Power Systems)

Widnall, William (engineer at MIT Instrumentation Lab)

Wood, William (systems engineer, Bendix/AlliedSignal, Jet Propulsion Laboratory, Goddard Space Flight Center)

Woodfill, Jerry (electrical engineer for the Apollo caution and warning system)

Wren, Robert (lead engineer for Apollo CSM and LM testing)

Young, Kenneth (aerospace technician, Lunar Rendezvous Section/Mission Analysis Branch)

Young, Larry (director, National Space Biomedical Research Institute)

JOHNSON SPACE CENTER ORAL HISTORIES

Arabian, Donald

Armstrong, Neil

Battin, Richard

Bond, Aleck

Borman, Frank

Chaffee, Norman

Cohen, Aaron

Collins, Michael

Deiterich, Charles

Faber, Stanley

Garman, John (Jack)

Griffin, Gerald

Heflin, Milton

Honeycutt, Jay

Hooks, Ivy

Hughes, Francis (Frank)

Johnson, Gary

Kelly, Thomas

Kraft, Christopher

Kranz, Eugene

Lee, Dorothy (Dottie)

Lee, John

Lunney, Glynn

McLane, James, Jr.

Mechelay, Joseph

Miller, Harold

Moser, Thomas

Osgood, Catherine

Pohl, Henry

Schweickart, Russell (Rusty)

Seamans, Robert

Shelley, Carl

Stafford, Thomas

Vaughan, Chester

Woodling, Carroll (Pete)

Wren, Robert

Young, Kenneth

BOOKS, JOURNALS AND OTHER DOCUMENTS

Aldrin, Edwin Eugene Jr. *Line-of-Sight Guidance Techniques for Manned Orbital Rendezvous*. Massachusetts Institute of Technology Department of Aeronautics and Astronautics, January 1963.

The Apollo Spacecraft: A Chronology. NASA, online publication.

Barbree, Jay. *Neil Armstrong: A Life of Flight*. New York: Thomas Dunne Books, 2014.

Benson, Charles D., and William Barnaby Faherty. *Moonport: A History of Apollo Launch Facilities and Operations*. NASA History Series. January 1, 1978.

Chaikin, Andrew. *A Man on the Moon*. New York: Viking Penguin, 1994

Compton, W. David. *Where No Man Has Gone Before: A History of Apollo Lunar Exploration Missions*. Dover Publications, 2010.

Cortright, Edgar M., ed. *Apollo: Expeditions to the Moon*. NASA History Office, 1975.

Dick, Steven J., and Roger D. Launius, eds. *Societal Impact of Spaceflight*. NASA History Series. Washington, DC: NASA, 2007.

Duffy, Robert A. *Biographical Memoirs: Charles Stark Draper*. National Academies Press, 1994.

Eyles, Don. *Sunburst and Luminary: An Apollo Memoir*. Boston: Fort Point Press, 2018.

Hacker, Barton C., and James M. Grimwood. *On the Shoulders of Titans: A History of Project Gemini*. Washington, DC: NASA, 1977.

Hansen, James R. *First Man: The Life of Neil A. Armstrong*. New York: Simon and Schuster, 2005.

Harland, David M. *The First Men on the Moon*. Berlin: Springer-Praxis Books, 2007.

Hoag, David G. *History of the Apollo On-Board Guidance, Navigation and Control*. September 1976.

James David Alexander Family. *The Moon and More*. N.p.: iUniverse, 2007.

Johnson, Gary. *Lessons Learned from 50+ Years in Human Spaceflight*. JSC SMA Flight Safety Office, April 30, 2018.

Lindsay, Hamish. *Tracking Apollo to the Moon*. London: Springer-Verlag, 2001.

Lunney, Glynn. *Highways into Space*. 2014.

Mangus, Susan, and William Larsen. *Lunar Receiving Laboratory Project History*. NASA/CR–2004–208938. NASA, 2004.

Mindell, David A. *Digital Apollo: Human and Machine in Spaceflight*. Cambridge, MA; MIT Press, 2008.

Murray, Charles, and Cathern Bly Cox. *Apollo: The Race to the Moon*. New York: Simon and Schuster, 1989.

Oates, Stephen B. "NASA's Manned Spacecraft Center at Houston, Texas." *Southwestern Historical Quarterly* 67, no. 3 (January 1964): 350–375.

Painter, John H., and George Hondros. *Unified S-Band Telecommunications Techniques for Apollo*. 2 vols. NASA Technical Notes. Washington, DC: NASA, March 1965, April 1966.

Seamans, Robert C. Jr. *Project Apollo: The Tough Decisions*. NASA Monograph Series. 2005.

Thimmesh, Catherine. *Team Moon: How 400,000 People Landed Apollo 11 on the Moon*. New York: Houghton Mifflin, 2006.

Tomayko, James. *Computers in Spaceflight: The NASA Experience*. NASA, 1988.

Walters, Lori C. *To Create Space on Earth: The Space Environment Simulation Laboratory and Project Apollo*. Houston: NASA, 2003.

Woodling, C. H., Stanley Faber, John J. Van Bockel, Charles C. Olasky, Wayne K. Williams, John L. C. Mire and James R. Homer. *Apollo Experience Report Simulation of Manned Space Flight for Crew Training*. NASA Technical Note TN D-7112. Washington, DC: NASA, March 1973.

Woods, W. David. *How Apollo Flew to the Moon*. New York: Springer-Praxis Books, 2011.

FILMS

Fairhead, David, dir. *Mission Control: The Unsung Heroes of Apollo*. Gravitas Ventures, 2017.

Moon Machines. Season 1, episode 3, "Navigation." Science Channel, 2008.

Sington, David, and Christopher Riley, dirs. *In the Shadow of the Moon*. Vertigo Films, 2007.

WEBSITES

APPEL News Staff. "A Strategic Decision: Lunar-Orbit Rendezvous." NASA. APPEL Knowledge Services, January 10, 2012. https://appel.nasa.gov/2012/01/10/5-1_lunar_orbit_rendezvous-html/.

Atkinson, Nancy. "How to Handle Moon Rocks and Lunar Bugs: A Personal History of Apollo's Lunar Receiving Lab." Universe Today, July 19, 2009. https://www.universetoday.com/35229/how-to-handle-moon-rocks-and-lunar-bugs-a-personal-history-of-apollos-lunar-receiving-lab/.

Atkinson, Nancy. "How We *Really* Watched Television from the Moon." Universe Today, August 7, 2009. https://www.universetoday.com/36950/how-we-really-watched-tv-from-the-moon/.

Bogo, Jennifer. "The Oral History of Apollo 11: The Knuckle-Biting Story of the First Lunar Landing from the People Who Were There." *Popular Mechanics*, July 17, 2018. https://www.popularmechanics.com/space/moon-mars/a4248/oral-history-apollo-11/.

CBS News Coverage of the Launch of Apollo 11, via YouTube.

Johnson Space Center History Collection online; JSC Roundup Archives.

Jones, Eric M., ed. Apollo Lunar Surface Journal. 1995–2018. https://www.hq.nasa.gov/alsj/.

Lunar and Planetory Institute website lpi.usra.edu, numerous pages.

Mackellar, Colin. "A Tribute to the Honeysuckle Creek Tracking Station, 1967–1981." December 2003–present.

NASA.gov., numerous pages.

Platoff, Anne M. "Where No Flag Has Gone Before: Political and Technical Aspects of Placing a Flag on the Moon." Johnson Space Center. NASA Contractor Report 188251, August 1993. https://www.jsc.nasa.gov/history/flag/flag.htm.

Pyle, Rod. "Apollo 11's Scariest Moments: Perils of the 1st Manned Moon Landing." Space.com, July 21, 2014. https://www.space.com/26593-apollo-11-moon-landing-scariest-moments.html.

Shira Teitel, Amy. "Apollo 11's '1202 Alarm' Explained." *Vintage Space* (blog). *Discover* magazine, January 5, 2018. http://blogs.discovermagazine.com/vintagespace/2018/01/05/apollo-11s-1202-alarm-explained/#.XGxKKOhKhPY.

Szondy, David. "Saturn V: The Birth of the Moon Rocket." New Atlas, July 4, 2018. https://newatlas.com/saturn-v-birth-moon-rocket/54867/.

Woods, David, Ken MacTaggart and Frank O'Brien. "Apollo 11 Flight Journal." NASA, March 1, 2016. https://history.nasa.gov/afj/ap11fj/index.html.

Facing page: Buzz Aldrin stands next to the Lunar Module during the Apollo 11 moonwalk as he prepares to deploy the Early Apollo Scientific Experiments Package (EASEP). Credit: NASA.

ACKNOWLEDGMENTS

MY SINCERE AND HEARTFELT THANKS to all the engineers and scientists who shared their unique Apollo stories and insights and who worked with me to ensure accuracy of technical and historical details. Although there wasn't space or breadth within this book to include all the tales I heard, everyone's stories helped provide such wonderful context for the amazing days of the 1960s.

Special thanks to the NASA Alumni League–JSC Chapter, especially "rocket scientist" Norman Chaffee, who arranged interviews and tours during my visit to Houston and provided constant encouragement; thanks to Milt Heflin, who put me in touch with the SimSups. Thanks to John Painter who provided early technical editing assistance. Thank you to Kay Ferrari of NASA's Solar System Ambassador program for putting me in touch with fellow ambassadors Earle Kyle and Rich Manley. My appreciation and thanks to media specialists Brandi Dean and Noah Michelsohn at Johnson Space Center, Daniel Dent at Draper and Sara Remus at MIT, as well as Dr. Sandra Johnson at the JSC History Center, who provided invaluable assistance.

Thank you to everyone who contributed photos for this book, with a special shout-out to Draper, Colin MacKellar from HoneysuckleCreek.net and of course, NASA.

My deepest thanks and appreciation to Page Street Publishing for the opportunity to write this book, with special thanks to my editor, Lauren Knowles, for her boundless enthusiasm and patience. Thank you to copy editor Nichole Kraft for her amazing and thorough work in smoothing out all the (numerous) rough spots in the manuscript. Cheers to my network of fellow journalists for your inspiration and support. Endless thank-yous and love to my incredible husband, Rick, for his support at every step, as well as to my family and friends for always providing such amazing help and encouragement and being interested in my quirky obsession with space exploration.

ABOUT THE AUTHOR

NANCY ATKINSON is a science journalist and author with a passion for telling the stories of people involved in space exploration and astronomy. Her first book, *Incredible Stories from Space: A Behind-the-Scenes Look at the Missions Changing Our View of the Cosmos*, tells the stories of thirty-seven scientists and engineers who work on several NASA robotic missions to explore the solar system and beyond.

Since 1999, Nancy has written thousands of articles on space and astronomy. She currently contributes to Universe Today and *Ad Astra*. Previously, she was an editor for Universe Today and editor in chief for *Space Lifestyle Magazine*. She has also written articles for Seeker, *New Scientist*, *Wired*, Space.com, NASA's *Astrobiology Magazine*, *Space Times* magazine and several newspapers in the Midwest. She has been involved with several space-related podcasts, including *Astronomy Cast* and *365 Days of Astronomy*, and was the host of the NASA Lunar Science Institute podcast. Nancy is also a NASA-JPL Solar System Ambassador. She lives in Minnesota.

INDEX